Advanced Computer
Performance Modeling
and Simulation

Advanced Computer Performance Modeling and Simulation

Edited by

Kallol Bagchi

Florida Atlantic University at Boca Raton

Jean Walrand

University of California at Berkeley

and

George W. Zobrist

University of Missouri at Rolla

CRC Press

Taylor & Francis Group

Boca Raton London New York

CRC Press is an imprint of the
Taylor & Francis Group, an **informa** business

CRC Press
Taylor & Francis Group
6000 Broken Sound Parkway NW, Suite 300
Boca Raton, FL 33487-2742

© 1998 by Taylor & Francis Group, LLC
CRC Press is an imprint of Taylor & Francis Group, an Informa business

No claim to original U.S. Government works

ISBN-13: 978-90-5699-569-0 (hbk)

Visit the Taylor & Francis Web site at
http://www.taylorandfrancis.com

and the CRC Press Web site at
http://www.crcpress.com

This book is dedicated to
all the people
who helped make it a success

CONTENTS

EDITORS' PREFACE

Performance modeling and simulation have been used in computer and communication system design for some time. This book makes the argument that this topic has become a central issue in computer science and engineering research and should be considered seriously. Its object is to lead researchers, practitioners and students involved in this discipline. Distinctive in many ways — it provides tutorials and surveys on important topics, relates new research results that should be of interest to all working in the field and covers a fairly broad area in the process. Each chapter presents background, describes and analyzes important works in the field and provides direction to the reader on future work and further readings. The volume can be used as a reference book by all associated with computer science engineering. Our hope is that this set of carefully selected papers will be of interest to computer science engineers, students, academicians and practitioners.

This is the first of four books concerning the state of the art in performance modeling and simulation of advanced computer systems and networks. It deals primarily with theory, tools and techniques as related to advanced computer system design. A soon-to-be-published accompanying volume describes specific models and simulators for such systems. In this book, the tools range from a hierarchical simulator to an object-oriented parallel system simulator to an instrumentation system for integrated parallel system design and evaluation. The theory ranges from Parallel Simulation to Petrinets to Stochastic Process Algebra (SPA). Novel techniques are described in performance evaluation as well as in simulation and modeling. The systems considered are mostly sequential, parallel (multiprocessors: shared and distributed memory) and distributed machines.

The book begins with a hierarchical architecture design simulation environment. Mr. Howell and Prof. Ibbett advocate a hierarchical design scheme with an integrated set of tools and describe such a tool set named HASE. Architectures can be designed at different levels of abstraction and the simulation can be viewed through animation of the design drawings. Hierarchical design schemes have been used in many applications and parallel system design tools like N.2, DORMS, and so on, and have been designed before based on a similar concept. (Please refer to *Euromicro Journal*, vol. 13, 1983 and vol. 16, 1985.) However, HASE is an advanced tool that uses animation facilities; it can be used for single- as well as multiprocessor-based design.

Chapters 2–5 deal with various aspects of parallel and distributed system modeling and simulation, especially multiprocessor systems. The second chapter is a tutorial on modeling of multiprocessor architectures. Prof. Malloy describes an execution-driven multiprocessor simulator and compares such simulation with trace-driven simulation. Trace-driven simulation is faster but requires more space. Execution-driven simulation, when properly designed, can make the verification process more straightforward. Execution-driven simulation is perhaps more costly to design and maintain, but once properly designed, it may be really helpful to researchers. Many such useful simulators were designed in the past and are routinely designed at present. For example, one of the first-named editor's students designed a Transputer T800 simulator using instruction-level simulation. (See Proceedings WMC'93, SCS Press.)

Integrated tool environments are needed for modeling, simulation and performance monitoring, especially in parallel and distributed system design. Chapter 3 is an effort in that direction and concerns a new system for parallel tools called Instrumentation System (IS). Mr. Waheed and Prof. Rover use the term IS to refer to "components that support runtime data collection and management activities in state-of-the-art tool environments for parallel and distributed systems." This chapter discusses the state-of-the-art in IS design and evaluates parallel and distributed tools according to various features of IS modules and services. In addition, a structured approach of designing and evaluating IS is discussed.

Tony Field and Peter Harrison detail a new methodology for the performance evaluation of distributed shared memory (cache coherent) multiprocessor systems (chapter 4). Dr. Field and Prof. Harrison discuss an efficient and general analytical modeling scheme for shared memory multiprocessor systems that can model both the architectures and coherency protocols. Various design changes can be accommodated easily in the model and model execution time is much lower than an execution-driven simulation. The authors show the practicality of their modeling scheme with two useful examples.

The fifth chapter discusses problems with tracing techniques related to shared memory multiprocessor simulation. Dr. Eckert and Prof. Nutt focus on the danger of relying too heavily on such event traces where the program's execution exhibits nondeterministic behavior. This is a new and important area being addressed increasingly by researchers. (Refer to similar works in the bibliography of this chapter and the MASCOTS '96 proceedings published by IEEE Press.)

Chapters 6 through 8 deal with parallel simulation tools and techniques —conservative and optimistic. Parallel simulation is crucial, as the computational needs of many simulation applications demand it; however, fast and efficient schemes are needed. Chapter 6 is a tutorial on memory management and speedup issues. Dr. Romdhane and Prof. Madisetti survey a number of recent techniques as related to Parallel Discrete Event Simulation (PDES) and deal with memory and speedup issues. Chapter 7 is concerned with new load balancing strategies on a multiprocessor machine. Profs. Boukerche and Das consider the problem of load balancing for parallel simulation on multiprocessor machines and experiment with several such techniques based on both conservative and optimistic approaches. Chapter 8 details an object-oriented environment. Profs. Bagrodia and Waldorf describe a system called MOOSE that is unique in several aspects: it separates model design and model execution; it allows a simulation model to be designed in an iterative manner from a prototype to parallel implementation; it is object-oriented and uses an inheritance mechanism to simplify parallel simulation design. Object-oriented efficient parallel simulation environments are rare. MOOSE is an effort in that direction.

Chapters 9 and 10 deal with theoretical models that can be used for formal as well as performance modeling. Chapter 9 is a tutorial on stochastic Petrinets. Prof. Ciardo reviews various classes of SPN with the latest techniques. SPNs have been used extensively in modeling various reliability and performance characteristics of computer and communication systems; there have been projects that used Petrinets to model parallel systems. Large-scale modeling with such nets is always a challenge to designers. Chapter 10 is a tutorial on SPAF. Prof. Hillston and Dr. Ribaudo discuss a new technique, Stochastic Process Algebra Formalism, based on stochastic timing and probabilistic branching. The authors state: "This novel approach to performance modeling, based on formally defined, compositional system descriptions, has benefits for model construction, model manipulation and model solution," and thus may help in making performance evaluation more formal and, at the same time, easy to grasp.

Chapters 11 and 12 concern performance evaluation techniques. Chapter 11 details performance evaluation using micro-benchmarking and machine analysis. Profs. Saavedra and Smith present a new methodology for CPU performance evaluation based on the concept of an abstract machine model and contrast it with benchmarking. By combining machine and program characterizations, accurate execution time predictions are obtained. A wide variety of computers are analyzed and results are presented that assert the usefulness of this new methodology. Chapter 12, on evaluation and design of benchmarking suites using quantitative methods, introduces new models that can be used for workload characterization, evaluation and design of benchmark suites. Prof. Dujmovic´ introduces a concept of program space coverage that opens up new directions in benchmarking research. He concludes that it is possible to design universal benchmark suites that can reduce the cost of industrial benchmarking substantially.

The challenge remains of significantly advancing the state of the art in this field as discussed above. The present articles show incremental changes. No doubt, in time, these incremental changes will appear ordinary and newer theories, tools and techniques will emerge. However, this book does capture many important aspects of developments in this field over the last ten years and will perhaps remain current for a number of years. If that happens, we will consider our efforts to be successful.

ACKNOWLEDGMENTS

This project was first conceived when I was working at Aalborg University in Denmark, in 1992 and 1993, as an associate professor of computer science and engineering. I "sounded out" some of my colleagues—experts and peers who agreed there was an acute need for this kind of book in the field. They promised help and cooperation in the successful completion of the project; and I began work on this and three companion books in the summer of 1994. Since then, many changes have taken place and, in the process, many debts of gratitude have been incurred.

Thanks are due to many people: first, my brother Dr. Milan Bagchi and his family; next, the Population Council of Rockefeller University in New York for allowing me to use their computer resources. Computer resources were utilized at STAR Lab, Stanford University, California and at Columbia University, New York. At Stanford, I used some of my time there as a visiting scholar in this project. I appreciate the tremendous support especially from the late Prof. Allan Peterson, Jim Burr and team of research scholars Gerard Yeh, Yen-wen Lu and Bevan Bass.

I am indebted to Pooran Rambharose of the Mechanical Engineering Department of Florida Atlantic University, Boca Raton, where I used the laser printer for some of the chapters. I am grateful to Carol Saunders and Drs. Rom Dattero, Paul Hart and Bob Cerveney of the DIS Department, at Florida Atlantic.

Dr. Carey Williamson of the University of Saskatchewan in Canada deserves special mention for providing help with LATEX script, which many authors used in preparing final versions of their chapters. Dr. Ana Pont Sanjuan of Spain merits thanks for lending others her Microsoft Word™ script for this purpose. Prof. Yi-Bing Lin of National Chiao Tung University, Taiwan, is greatly appreciated for providing authors help in generating index terms using LATEX. Many others provided feedback for improving the structure of the book; thanks are due to all of them.

Profs. Kishor Trivedi and Erol Gelenbe, both at Duke University, Durham, North Carolina, provided a great deal of friendly advice.

The editors wish to thank all the editorial board members and contributors, without whose patience and cooperation this book would not have been possible. Acknowledgment is also due many people who provided anonymous support in different ways. In brief, it seems that we have too many debts of gratitude to recount. So we take this opportunity to thank all who stood by us during various phases of this project.

Kallol Bagchi

CONTRIBUTORS

Rajive Bagrodia, Computer Science Department, 3532 Boelter Hall, Box 951596, University of California, Los Angeles, California 90095 USA email: rajive@cs.ucla.edu

Azzedine Boukerche, Department of Computer Science, University of North Texas, P.O. Box 13886, Denton, Texas 76203-6886 USA email: azedine@mireille.cs.unt.edu

Gianfranco Ciardo, Department of Computer Science, Tercentenary Hall, College of William and Mary, P.O. Box 8795, Williamsburg, Virginia 23187-8795 USA email: ciardo@cs.wm.edu

Sajal K. Das, Department of Computer Science, University of North Texas, P.O. Box 13886, Denton, Texas 76203-6886 USA email: das@cs.unt.edu

Jozo J. Dujmović, Department of Computer Science, San Francisco State University, 1600 Holloway Avenue, San Francisco, California 94132 USA email: jozo@sfsu.edu

Zulah K. F. Eckert, Department of Computer Science, CB 430 University of Colorado, Boulder, Colorado 80309-0430 USA email: eckert@cs.colorado.edu

A. J. (Tony) Field, Department of Computing, Imperial College of Science, Technology and Medicine, 180 Queen's Gate, London SW7 2BZ United Kingdom email: ajf@doc.ic.ac.uk

Peter G. Harrison, Department of Computing, Imperial College of Science, Technology and Medicine, 180 Queen's Gate, London SW7 2BZ United Kingdom email: pgh@doc.ic.ac.uk

Jane E. Hillston, Department of Computer Science, University of Edinburgh, The King's Buildings, Mayfield Road, Edinburgh, Scotland EH9 3JZ United Kingdom email: jeh@dcs.ed.ac.uk

Frederick W. Howell, Department of Computer Science, University of Edinburgh; The King's Buildings, Mayfield Road, Edinburgh, Scotland EH9 3JZ, United Kingdom email: fwh@dcs.ac.uk

Roland N. Ibbett, Department of Computer Science, University of Edinburgh, The King's Buildings, Mayfield Road, Edinburgh, Scotland EH9 3JZ, United Kingdom

Vijay K. Madisetti, School of Electrical Engineering, DSP Lab, Georgia Tech, Atlanta, Georgia 30332 USA email: vkm@eedsp.gatech.edu

Brian A. Malloy, Department of Computer Science, Clemson University, 416 Edwards Hall, Clemson, South Carolina 29631-1906 USA email: malloy@cs.clemson.edu

Gary J. Nutt, Department of Computer Science, CB 430 University of Colorado, Boulder, Colorado 80309-0430 USA email: nutt@cs.colorado.edu

Marina Ribaudo, Dipartimento di Informatica, Universita di Torino, corso Svizzera 185, 10149 Torino, Italy email: marina@di.unito.it

Mohamed S. Ben Romdhane, Science Center, Rockwell Intl. Corporation, 1049 Camino Dos Rios, Thousand Oaks, California 91360 USA email: romdhane@jupiter.risc.rockwell.com

Diane T. Rover, Department of Electrical Engineering, 260 Engineering Building, Michigan State University, East Lansing, Michigan 48824 USA email: rover@ee.msu.edu

Rafael H. Saavedra, Computer Science Department, Henry Salvatory Computer Science Center, University of Southern California, University Park, Los Angeles, California 90089-0781 USA email: saavedra@catarina.usc.edu

Alan Jay Smith, Computer Science Division, University of California, Berkeley, California 94720-1776 USA email: smith@ylem.berkeley.edu

Abdul Waheed, Department of Electrical Engineering, 260 Engineering Building, Michigan State University, East Lansing, Michigan 48824 USA email: waheedr@egr.msu.edu

Jeff Waldorf, Software Technologies Corporation, Los Angeles, California USA

Editorial Board

Jan Barr, STAR Lab, Stanford University and SUN Microsystems, Palo Alto, California USA email: barr@nova.stanford.edu

Predip K. Das, Department of Computer Science and Engineering, Jadavpur University, Calcutta 700 032, India

Patrick Dewit, Department of Electrical Engineering, University of Maryland, A.V. Williams Bldg., College Park, Maryland 20742 USA, email: dewit@eng.umd.edu

Aura J. Dutigsmid and Vijay R. Madisetti — See List of Contributors

Editors

Kartali dapola, Department of DIS, Florida Atlantic University, Boca Raton, Florida 33431 USA email: abea30@cse.fau.edu

Jean Walrand, Department of Electrical Engineering, University of California, Berkeley, California 94720 USA email: walr@eecs.berkeley.edu

George W. Zobrist, Department of Computer Science, University of Missouri, 1870 Miner Circle, Rolla, Missouri 65409-0350 USA, email: zobrist@umr.edu

CHAPTER 1

HIERARCHICAL ARCHITECTURE SIMULATION ENVIRONMENT

F.W. Howell and R.N. Ibbett

1.1. INTRODUCTION

The Hierarchical Architecture Simulation Environment (HASE) is a tool for modelling and simulating computer architectures. Using HASE, designers can create and explore architectural designs at different levels of abstraction through a graphical interface based on X-Windows/Motif and can view the results of the simulation through animation of the design drawings. This chapter describes the design and animation facilities of HASE, compares it with other simulation systems and concludes with suggestions for future tools based on several years' experience using HASE within the University of Edinburgh department of computer science.

1.1.1. The Motivation

Advanced simulation tools are available for low level electronic design, such as Spice for analogue circuits, and VLSI layout tools. However, tools for rapid prototyping of architectural ideas are less well established. Simulation languages can be used to model computer architectures, but the user has to be an expert on simulation. This is also the problem of general purpose simulation tools (e.g. SES/Workbench), where icons represent 'queues', 'servers' etc., and the link between a queueing model of an architecture and the architecture itself is not immediately apparent to the engineer not fluent in queueing theory.

1

Conventional languages (C, C++) are often used to construct simulators, but this approach involves starting from scratch for each new project. User interface aspects are often neglected as the tool will be thrown away with the next architecture. This is very wasteful, as many aspects of computers are constant between different architectures. The object oriented approach offers a solution. Standard components (such as memories, microprocessors and interconnection networks) can be held in a library. They can be constructed and linked together graphically on screen to create a simulation of an architecture, in much the same way that standard components can be wired together in a semi-custom VLSI tool. The difference is that the simulation is not fixed to low level wires and chip pins, but is free to choose the appropriate abstraction level.

HASE was designed to provide the flexibility of a raw programming language with the user interface advantages of a graphical tool.

1.2.   DESIGN OF HASE

1.2.1.   Overall operation

The HASE tool acts as a graphical front end to SIM++[1], a discrete event simulation extension of C++. SIM++ is used to describe the behaviour of basic components of a simulation. It provides a sim_entity class from which user components may be derived. Entities run in parallel and may schedule messages to other entities using SIM++ library functions. The user can link icons corresponding to entities together on screen and HASE produces the SIM++ initialisation code necessary for simulating the network. New components can be constructed by linking together standard components. Each component can be simulated at any level of abstraction. A register transfer level simulation will produce the most accurate simulation results; behavioural level simulations run more swiftly. The tool allows different parts of the simulation to run at different abstraction levels, so the user can 'zoom in' to specific parts of the design to simulate that at a low abstraction level and run the rest of the design at a high level of abstraction. Figure 1.1 shows how the parts of the system fit together.

1.2.2.   Internal design of Hase

Each project built using HASE has its own directory for storing the SIM++ code. This directory may be used for building and running the simulation

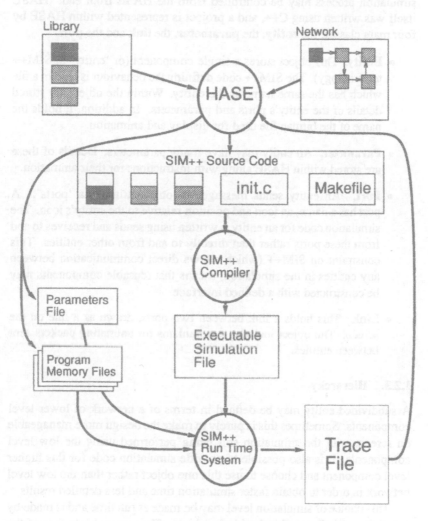

Figure 1.1. The top level design of HASE.

outwith the HASE environment using command line tools like make, giving
the full flexibility of the SIM++ programming language. Alternatively the
simulation process may be controlled from the HASE front end. HASE
itself was written using C++, and a project is represented within HASE by
four main classes; the **entity**, the **parameter**, the **link** and the **port**.

- **Entity.** This object stores a single component (or 'entity' in SIM++
  terminology). The SIM++ code defining the behaviour is held in a file
  which has the same name as the entity. Within the object are stored
  details of the entity's ports and parameters. In addition, it holds the
  name of the bitmap file used for display and animation.

- **Parameter.** An entity may have many parameters. Details of these
  are stored within HASE along with instructions for their animation.

- **Port.** An entity sends messages to other entities via 'ports'. A
  port has a name, an icon and position relative to the entity's icon. The
  simulation code for an entity is written using sends and receives to and
  from these ports rather than directly to and from other entities. This
  constraint on SIM++ (which allows direct communication between
  any entities in the simulation) means that reusable components may
  be constructed with a defined interface.

- **Link.** This holds a link between two ports, drawn as a line on the
  screen. The object includes mechanisms for animating packets sent
  between entities.

1.2.3. Hierarchy

A subdivided entity may be defined in terms of a network of lower level
components. Sometimes this is purely to make the design more manageable
on screen, with the simulation still being performed using the low level
components. It is also possible to provide simulation code for this higher
level component and choose to use this one object rather than the low level
network in order to obtain faster simulation time and less detailed results.

This choice of simulation level may be made at run time and is made by
toggling a switch associated with the object. The external interface of the
high level component is defined to be the same as that of the lower level
network. This allows the simulation level of each object in the simulation
to be set independently. Figure 1.2 illustrates two subdivided components
connected by their external ports.

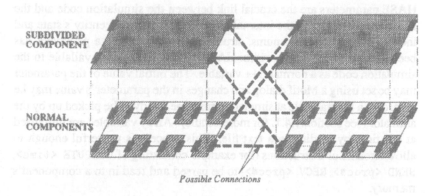

Figure 1.2. Two subdivided entities are connected by their external ports.

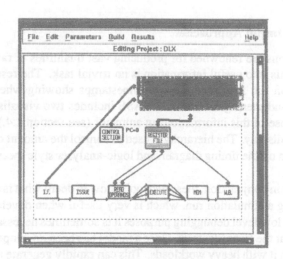

Figure 1.3. The HASE user interface.

1.2.4.   Parameter Types

HASE parameters are the crucial link between the simulation code and the animation. They form the internal representation of each entity's state and include integers, floats, enums, structs and arrays. Once a parameter has been defined for an entity within HASE, that parameter is available to the simulation code as a normal C++ variable. The initial value of the parameter may be set using a Motif dialog and changes in the parameter's value may be recorded in the trace file at simulation run time, ready to be picked up by the animator (see section 1.4.1. for more details). Array variables are initialised at run time by reading in a text file. This process is powerful enough to allow streams of instructions (for example consisting of COMPUTE <time>, SEND <proc#>, RECV <proc#>) to be parsed and read in to a component's memory.

1.2.5.   Templates

Templates for building common structures such as arrays and meshes of components are included. The user can slot any component into the template, set the dimensions and all the required components and links are produced. Current templates include a linear array, a 2D mesh, an omega network and a 3D torus.

1.2.6.   Output Approaches

Simulations are renowned for producing vast quantities of raw data; transferring this into useful information is no trivial task. The result of a single simulation run is a trace file with timestamps showing when all changes in state and messages occurred. HASE includes two visualisation tools to make sense of this information; an animator (see section 1.4.) and a timing diagram display. The hierarchy is used to control the amount of information displayed on the timing diagram and logic-analyser style measurements can be taken.

   Used in conjunction, these two tools show in detail what is actually going on during a simulation run, which is very useful when developing models. For very low level debugging purposes it is sometimes necessary to resort to looking at the trace file itself. Once a model has been developed, it is natural to stretch it with heavy workloads. This can rapidly generate unmanageably large trace files, so there is a mechanism in Hase for controlling how much trace information is produced (section 1.4.1.). For the largest runs it is usual to garner a small number of statistical measures from the model. These measures are taken using classes provided in SIM++ for histograms, counts and accumulated averages.

Repeated runs are required to investigate how a model behaves using a range of parameters[2]. These runs are typically controlled by a Perl script and graphs produced using the GNUplot program.

1.2.7. Recycling Simulation Objects

One of the major benefits ascribed to object oriented techniques is that software components may be reused by others instead of being recreated from scratch.

This ideal has nearly been attained by hardware simulation systems; hardware components have well defined inputs and outputs so designs may be constructed by gluing together off-the-shelf components. The ideal is only "nearly" attained in this case as effort is still required to package components for others to use, so a certain amount of reinvention still occurs.

The situation isn't so rosy with object oriented software. This is partly because software is inherently more flexible than hardware. It becomes more difficult to define interfaces between objects when they aren't constrained to N physical wires, but may instead be composed of data types, interdependent methods, global variables and so on. It requires a significant investment in time and effort to document and prepare objects so others may use them[3]. As a result, few objects are generally shared between people, and most people only reuse code they have written themselves.

It was an early design aim of the Hase system to encourage object re-use as much as possible. This has met with some success in practice (but not as much as was hoped for). The interface to most Hase objects is by typed messages to ports, which makes reuse of objects simpler than the general C++ case (but not quite as straightforward as low level hardware models). Objects which play by these rules may be included in a project with no problems. However Hase does not enforce this model; it is possible for objects to use SIM++ techniques to communicate using global variables or to bypass the ports. This makes it more complicated to simply slot such an object into a project. Practicalities such as proper documentation being provided for objects also affect reuse.

The Hase library system has been designed to address these issues. Rather than storing a set of *components*, it stores a set of *projects* each of which includes a list of components, the parameter and message type declarations and the global variables.

1.2.8. Object Oriented Databases

There has been substantial commercial and academic interest in object oriented databases recently. One common type of object oriented database is an extension to an object oriented language (such as C++) which provides for *persistence* of the objects. This approach is advertised as being suitable for storing the complex objects common in CAD systems, and providing desirable facilities such as version control and checkpointing of designs.

To investigate this approach to managing designs, Hase objects were made persistent by using the ObjectStore[4] database system. The experience was not without its problems. All HASE source files had to be preprocessed by the ObjectStore compiler before seeing the C++ compiler, which lengthened compile times. General run time performance became sluggish as all standard C++ pointers were replaced with persistent pointers, which could potentially result in a disk access. Any changes to class definitions made all previous database files unreadable (unless they were processed using a command line tool). Substantial source code modifications were required to be compatible with ObjectStore assumptions, and more modifications were later needed to obtain reasonable performance.

The conclusion from this experiment with object oriented databases is that the technology isn't yet mature enough for this type of CAD system. The general idea of allowing persistent objects within a language (without requiring I/O code) is a good one to be greeted with enthusiasm; in practice, however, adding an object oriented database requires much more effort than it would take just to write I/O code.

1.2.9. Limitations of graphical simulation systems

Die hard hackers sneer at graphical tools in general since they may never be as flexible as a programming language. This lack of flexibility is indeed a problem with *entirely* graphical tools which construct models at all levels by joining icons. At the lowest level of design, a description in a programming language is often best. However, there are also limitations with *entirely* textual descriptions; hardware and software designers usually use pictures to explain a system in terms of its subsystems. A compromise is therefore in order.

HASE is an inherently graphical system; if no pictures are needed, then there is little point in using it. However it does not impose a graphical approach to the specification of individual objects. These are described in SIM++ and the full power of SIM++ is available to the programmer.

This compromise is finely balanced and it typically changes during the life cycle of a simulation project. Initially when the design is fluid, animation and graphics are very important for communicating ideas between researchers. Later, when the design solidifies, the important aspect is simulation run times for collecting experimental data.

1.3. OTHER APPROACHES

1.3.1. VHDL

VHDL has become established as the standard hardware simulation language. It enjoys support from all major EDA companies and provides for simulation at levels from behavioural down to gate level. This section compares the VHDL approach with using a C++ based simulation language for simulating hardware systems.

1.3.1.1. Why use anything other than VHDL?   High level simulations incorporating software are usually written in C or C++ since these are the languages used by programmers. It is possible to link code from different languages, but the process is never entirely painless as interface routines have to be written to convert between the different data formats. The ideal is to use one language throughout. McHenry[6] uses VHDL for high level system modelling, and Swamy[7] describes object-oriented extensions to VHDL to make it more suited to system modelling.

VHDL incorporates very powerful features for modelling hardware; there are explicit constructs for wires signals and detailed timing information may be included. It's possible to detect glitches and other low level hardware problems.

At the software level, good support is also included for concurrent processes; e.g.

```
architecture behavioural of component is
  signal w : bit := '0';
begin
  proc1: process is
  begin
    w <= 1;
```

```
        wait for 10 ns;
        w <= 0;
        wait for 10 ns;
      end;
      proc2: process is
      begin
        wait until w = '0';
      end;
    end behavioural;
```

Concurrent processes may be included *within* the description of a component. In SIM++, the unit of concurrency is the *entity* object. These entities communicate by sending and receiving *events*, which may contain data objects themselves. There is no concept of a *wire* as there is in VHDL, and no concept of a hierarchy of components (all entities are equal and may send messages to any other entity). The hierarchy is imposed on SIM++ by the Hase concept of ports. Programming in SIM++ is akin to programming a message passing parallel program.

The primary advantage of C++ based simulation languages (such as SIM++) over VHDL for system simulation is that linking to software libraries is significantly more straightforward. Basing communication upon messages passed between components rather than upon asserting signals allows a higher level view of the system, with the ability to send a data object at any abstraction level. VHDL on the other hand has much better tool support and standardisation than the various C++ simulation systems and includes direct support for modelling low level wire behaviour.

1.3.2. SIMULA / DEMOS

Another popular simulation approach is based on SIMULA and the discrete event package built on top of it (DEMOS). The original version of HASE was based on DEMOS[5]; the switch to SIM++ was motivated by the higher performance of C++ and the desire to interface to existing C and C++ libraries of code. Interaction between objects is based on shared *resources* which may have several operations defined, such as wait, coopt (a synchronisation).

1.3.3. Ptolemy

The Ptolemy project at Berkeley is a wide ranging simulation effort with a focus on signal processing[8]. It is a framework encompassing many different simulation styles, including a discrete event domain. The package includes support for animations written manually using the Tcl/Tk toolkit.

1.3.4. Commercial Tools

Several commercial tools are available for network modelling and general system simulation, an example being BoNeS[9]. These tools present a slicker and more complete interface than research prototypes like HASE, but as their source code isn't freely available they are less suited to playing with new ideas and adding new features.

1.4. ANIMATION

Watch the cogs and pistons of a steam engine for a while and you get a feel for the workings of the machine. This is trickier with electronic systems; although they are many times more complex than the steam engine, they just appear to sit and work their magic without effort (bar the odd flashing light and smoldering component).

An animation of a simulation model can generate a similar intuitive feel for how an electronic machine works. This often suggests 'obvious' improvements and highlights design flaws which may be concealed by a flat diagram or descriptive paragraph. It is also fun to watch a complex design coming alive on screen and behaving as intended (or, as is more likely, *not* behaving exactly as intended).

The main reason that animation isn't usually an integral part of the design process is the amount of effort involved in building one. The problem with creating an animation separate from the main design is that changes to the design have to be made to the animation code as well. This makes the animation diverge from the actual design and become unusable.

Hase addresses this by making animation an *integral* part of design. Simple animations are generated automatically, based on the state changes of components and the messages which are passed between them. More complex animations may be customised to include GIF colour icons.

1.4.1. The Approach

Animation is based on the changes in value of a component's parameters. These may be dragged onto the screen using the component editor (figure 1.4); once this has been done, any time that parameter's value changes it appears on the display.

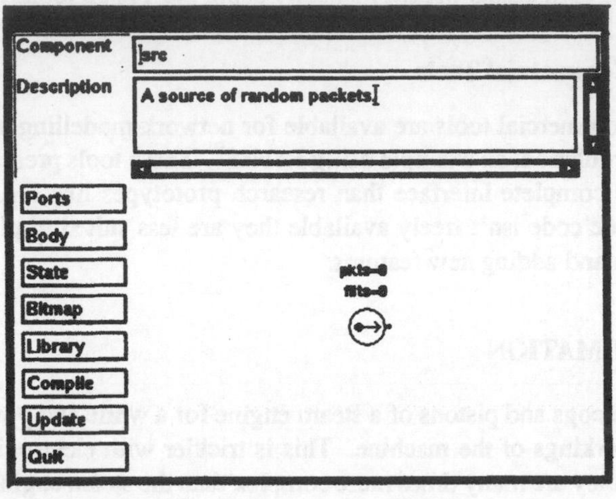

Figure 1.4. The component editor allows the state variables of an object to be dragged onto the display for animation.

The way a parameter is shown may be varied. Value just shows the value in screen (e.g. 123 for integers, 1.234 for floats, BUSY for enums). Name+Value shows the variable name as well (e.g. curr_state = BUSY).

Enumerated parameters may be displayed as icons instead of text; the icons are read in from bitmap files with the same name as the state (e.g. BUSY.btm or IDLE.gif). This is a simple but powerful technique for state animations; by simply providing the bitmaps for the corresponding states a customised animation is generated. These bitmaps may be displayed alongside the entity, or alternatively may be used to set the entity's bitmap.

struct parameters are displayed by drawing a box around the constituent elements (each of which may be displayed as above).

Thus far attention has been focussed on animating single parameters; any number of a component's parameters may be dragged onto the screen to be shown during animation, or they may be left hidden. It is also possible to define array parameters. The contents of these may be displayed on screen in a list box with a scroll bar and any updates or reads from the array are highlighted during the animation. Such updates are written to the trace using the MEM_READ() and MEM_UPDATE() macros in the SIM++ code. This technique has proved useful for displaying register contents and instruction buffers.

A simulation is not solely composed of state changes; there are also the messages sent between components. These messages may contain any form of data or handshake signal. The basic icon for a "message" may take any of the forms of static state parameters outlined above. This icon is animated by moving it down a link from one entity to another. The requisite line in the trace file is generated by the send_DATA() function in the SIM++ code, and the animation of the message is performed *at the time the message is sent*. Note that this is not necessarily the same as the time the message is acted upon by the receiving entity, as every SIM++ message is queued until the receiver is ready for it.

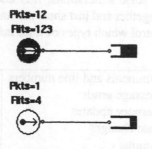

Figure 1.5. Changes in a component's state may be displayed on screen.

To show how the simulation code relates to the animation, figure 1.5 shows a src object connected to a queue and the following fragment shows part of the corresponding SIM++ code .

```
// excerpt from src.sim
    Pkts++;
    Flits++;
    if (ok_to_send)
       state=SRC_OK;
    else
       state=SRC_BLOCKED;
    dump_state();
    DataPkt d(123);
    sim_hold(1.234);
    send_DATAPKT(out,d,0.0);
    sim_wait(ev);
```

An example shows the format of the trace file which is generated on running the simulation and read in by the animator:-

```
// example trace file generated at run time
u:src0   at 0.000: P SRC_BLOCKED 12 123
u:queue0 at 0.000: P FULL_6
u:src1   at 0.000: P SRC_OK 1 4
u:queue1 at 0.000: P FULL_1
u:src0   at 1.234: S out 123
```

Sometimes protocols require several messages to be exchanged between entities; in these cases it would be messy to animate all the acknowledge packets, so it is possible to send messages without generating any trace information. For large scale simulations, it is also often useful to avoid animating messages altogether and just show the state changes, so the "trace level" may be set to control which types of trace information are generated. The levels are:

comments and line numbers	1
message sends	2
memory updates	3
state changes	4
summary	5

Table 1.1. The levels of trace generation.

Setting the trace level to 4 (say) includes state updates and summary information in the trace, but not messages, memory updates or comments.

1.4.2.   An example

Figure 1.6 shows an animation of a crossbar interconnection network with input and output queues. When the inputs block the icon is highlighted; it is possible to see the individual flits moving down the links and the queues grow and shrink dynamically.

1.5.   APPLICATIONS

Architectural simulation work using the DEMOS prototype version of HASE is detailed in[5]. In 1992 work began on the current SIM++/Motif version which has been used in many MSc and final year honours projects, including simulation of multiprocessor WAN bridger/routers, simulation of the DLX processor and simulation of the DASH multiprocessor[10]. More details of

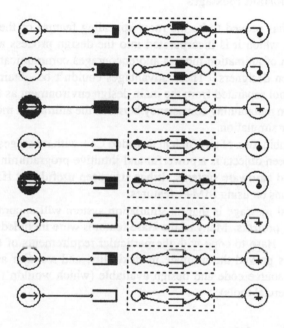

Figure 1.6. An interconnection network with input and output queues demonstrates the HASE animation facilities.

projects using Hase are given in[11]. Currently the main focus is on simulating multiprocessor interconnection networks and parallel MPI software. Many of the projects have involved linking simulation code to substantial existing libraries of C or C++ routines.

1.6. CONCLUSIONS

1.6.1. Important Messages

Animation has proved to be the most appealing feature of the Hase tool. The way in which it is incorporated into the design process allows swift construction of animation models and encourages communication and debate between designers. These advantages couldn't be obtained with an animation tool separated from the main design environment as there would be a problem maintaining consistency between the animation model and the one used for simulation.

The combination of an efficient threaded C++ with messages to communicate between objects is a powerful and intuitive programming model for software and hardware systems. It has also been useful that Hase imposes no restrictions on using SIM++ features.

The final message is that no simulation system will encompass all the needs of all projects. Many of the Hase features were included by students "extending" Hase to cope with the particular requirements of their project and this has proved the ultimate in flexibility, and a major advantage of having the source code and design available (which wouldn't be the case with commercial tools).

1.6.2. Future of the approach

New directions for the tool currently being investigated are closer tie-ins with an object oriented version of VHDL (to strengthen the links with hardware). VHDL itself is an attractive language for modelling hardware, but needs the addition of messages to model systems at a higher level. For software systems, it is very convenient to use a C/C++ like language since this makes it easy to include existing libraries of software.

Use of a parallel simulation language has been considered since the start of the Hase project and SIM++ originally had a timewarp version, but in projects to date the bottleneck hasn't been the simulation run time of individual runs, but rather the time to construct simulations. The lengthy simulations have been successive runs with different parameters which have been run simultaneously on different workstations. We are currently experimenting with our own implementation of SIM++ to run on the Cray T3D to map out the performance of a model over a large area of the input parameter space in parallel.

REFERENCES

1. JADE INC, Sim++ User Manual, (Jade Simulations International Corp., Calgary, Canada, 1992).

2. J. HILLSTON, A Tool To Enhance Model Exploitation, Technical Report CSR-20-92, Dept. of Computer Science, University of Edinburgh, 1992.

3. B. STROUSTRUP, The C++ Programming Language (Addison-Wesley, 1991), 382-384.

4. OBJECT DESIGN INC, ObjectStore Release 3.0 User Guide, (Object Design Incorporated, Burlington, MA, 1993).

5. A.R. ROBERTSON and R.N. IBBETT, "HASE: A Flexible High Performance Architecture Simulator", in Proc HICSS-27 (IEEE, Hawaii, 1994).

6. J.T. McHENRY and S.F.MIDKIFF, "VHDL Modeling for the Performance Evaluation of Multicomputer Networks", in Proc MASCOTS-94, (IEEE Computer Society Press, New York, 1994).

7. S. SWAMY, A. MOLIN and B. COVNOT, "OO-VHDL: Object-Oriented Extensions to VHDL", IEEE Computer, **28:10**, 18-26 (1995).

8. J. BUCK, S. HA, E.A. LEE and D.G. MESSERSCHMITT, "Ptolemy: A Framework for Simulating and Prototyping Heterogeneous Systems", Int. J. Comp. Sim., **4**, 155-182 (1994).

9. S.J. SCHAFFER and W.W. LaRUE, "BONeS DESIGNER: A Graphical Environment for Discrete-Event Modelling and Simulation", in Proc MASCOTS-94, (IEEE Computer Society Press, New York, 1994).

10. L.M. WILLIAMS, Simulating DASH in HASE, (MSc Dissertation, Department of Computer Science, University of Edinburgh, 1995).

11. R.N. IBBETT, P.E. HEYWOOD and F.W. HOWELL, "HASE: A Flexible Toolset for Computer Architects", to appear in The Computer Journal, (1996).

CHAPTER 2

MODELING MULTIPROCESSOR ARCHITECTURES

Brian A. Malloy

2.1. INTRODUCTION

Rapid technological developments have increased computational speed
and power through the exploitation of parallelism. However, it is often difficult to test new developments for parallel architectures since
the architecture may not be available or may be too expensive for the
researcher to obtain. A common approach to solving the problem of
unavailability of the target architecture is to use a simulator to capture
the behavior of the architectural system under development.

There are three common approaches to the design of a simulator:
the event-based approach, the time-based approach and the process-
based approach. Each of these approaches can be distinguished by
the method used to control the passage of time. In the event-based
approach, time is dynamically updated to reflect the time of the next
scheduled event in the model. In the time-based approach, a list of
activities is examined on each timing cycle to determine if a state change
can occur; time is updated at the end of each cycle. In the process-based
approach, the actions of each object in the model are incorporated into
a corresponding process that is then scheduled for execution; time in the
process-based approach is updated to reflect the time of the currently
active process in the model. The approach chosen, when designing
a model, should be based on the characteristics of the system under
study. For example, having time progress in ticks may provide more
control when simulating a developing architecture where events happen
regularly (every cycle) and in fixed increments[15].

One accepted approach to simulating an existing or developing architecture is *trace-driven* simulation, where the simulation is based on
a predetermined instruction sequence or *trace*. Trace-driven simulation

requires information about branch targets and memory references made during the execution of the program. This technique is used extensively to evaluate uniprocessor cache performance and techniques have been developed for capturing trace information to include references from system calls and the operating system.

Trace-driven simulation has been used to analyze multiprocessors[4]. However, there are several problems with applying the trace-driven approach to multiprocessor simulation. The simulated architecture must be similar to the architecture on which the trace was obtained, a significant problem when simulating an architecture that is in the development stage[7]. For a new architecture, the order of events, the latency of the network and the number of processors might be completely different from the trace host, rendering the traces useless when evaluating timing-sensitive applications. Using traces obtained from a sequential system may be inappropriate for use in simulating a parallel system since the order of execution is usually dependent on the order in which different parts of a parallel program complete execution[13]. Previous research has shown that traces obtained on a multiprocessor can produce erroneous results when used to simulate a different multiprocessor[3].

In this chapter, we present *execution-driven* simulation, an alternative to trace-driven simulation. Execution-driven simulation is also referred to as instruction-level simulation, register-transfer simulation or cycle-by-cycle simulation. We present the design of an execution-driven multiprocessor simulator and we compare the execution-driven approach to the trace-driven approach. In the execution-driven approach, the simulator actually executes an input program, allowing the development of software together with the hardware. Our simulator is coded in the process-based simulation language SimCal[12], and is guided by a *parameterized computational model* that describes the target multiprocessor. By varying the parameters to the model, we are able to vary the architectural configuration of the target multiprocessor. These variations include the number of processors, the parallelism in the interconnection network, the speed of the instruction set and the speed of the synchronization primitives. We use the multiprocessor simulator to execute parallel threads produced by our parallelizing compiler[10, 11] and we show, through the simulations, that our compiler is able to produce significant speed-up in the parallel execution of the program. Through experimentation, we derive parameters that accurately describe the AT&T 3B2/1000 multiprocessor. We then execute the threads on the 3B2, and observe a correlation in the speed-up achieved on the 3B2 compared to the speed-up achieved in the parameterized simulations. This correlation establishes confidence that our simulator can predict, for a given sequential program, the performance of a parallelized version of the program on an actual multiprocessor.

To compare the execution-driven and trace-driven approaches, we instrument the execution-driven simulator to construct traces of the program execution. Previous approaches to obtaining traces typically instrument the object code of the program which can induce perturbation of the timing results. Using the computational model and the traces obtained from the execution-driven approach, we perform trace-driven simulation and obtain the same speed-up in the trace-driven approach that we obtained in the execution-driven approach. A key facet of our work is the design of the computational model, which we incorporate into both the trace-driven and the execution-driven approach. In comparing the trace-driven approach with the execution-driven approach, we observe that the trace-driven approach is faster than the execution-driven approach since it does not actually execute the threads. However, the trace-driven approach must store the traces and therefore requires more space than the execution-driven approach. Finally, since the execution-driven simulator actually executes the input program, the output from the simulated parallel execution can be compared to the output of the original sequential program; this comparison is a functional verification of the threads produced by the parallelizing compiler.

The next section presents an overview of our approach to comparing trace-driven and execution-driven simulation and Section 2.3 presents our parameterized computational model. In Section 2.4, we describe our execution-driven simulator and in Section 2.5 we compare speed-up achieved using the simulator with speed-up achieved on an actual multiprocessor. We draw conclusions in Section 2.6.

2.2. OVERVIEW

Our approach to comparing trace-driven and execution-driven simulations is summarized in Figure 2.1. At the top of Figure 2.1, we indicate that the input to our parallelizing compiler[10, 11] is a *Source Program*, a general purpose program written in an imperative language. The *Par-*

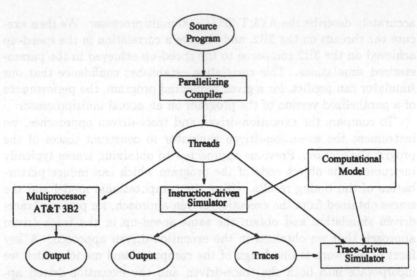

Figure 2.1: Approach summary. This figure summarizes the overall approach to comparing trace-driven and execution-driven simulations.

allelizing Compiler automatically parallelizes the sequential program by constructing threads that consist of statements from the source program and synchronization primitives that preserve the control and data dependences in the original program.

As Figure 2.1 illustrates, the *threads* are used as input to 1) the actual *Multiprocessor* (the AT&T 3B2/1000), 2) the *Execution-driven Simulator*, and 3) the *Trace-driven Simulator*. In Section 2.5, the speed-up achieved in the actual execution on the 3B2 is compared to the speed-up achieved by the execution-driven simulator. This comparison indicates that the simulations are a good predictor of the speed-up achievable in the actual executions. The *Output* from the 3B2 multiprocessor is compared to the *Output* of the execution-driven simulations; since the output is the same in both cases, this comparison serves to functionally validate the correctness of the parallelizing compiler in constructing the threads. There is no output from the trace-driven simulations since the instructions in the threads are not executed.

A goal of this work is to provide a valid platform for comparing the trace-driven and execution-driven approaches. For simulations that

Source program:	Thread for cpu_1	Thread for cpu_2
I := 0;	I:= 0;	
S := 0;	L2: **test** I ≤ N	S := 0;
while I ≤ N **loop**	**barrier**	L2: **barrier**
I := I + 1;	**brif** L1	**brif** L1
S := S + 1;	I := I + 1;	**Rv** I
end loop	**Sd** I	S := S + 1;
	branch L2;	**branch** L2;
	L1:	L1:

Figure 2.2: Threads example. This figure illustrates a source program, and two possible threads for the source program.

are sensitive to timing results, such as measuring speed-up achieved in parallelization, it is essential that the trace gathering process does not perturb the timings in program execution. However, previous research has established that it is nearly impossible to collect traces without perturbing program execution in some way[7]. Furthermore, it has been shown that results obtained from trace-driven simulation methods are not valid for evaluating multiprocessor performance[3]. Thus, we gather our traces from the execution-driven simulation where we can eliminate perturbation in program execution. Our host machine for the simulations is the same for both execution-driven and trace-driven simulations: an IBM PS/2 with an Intel 80486/SX processor. As Figure 2.1 indicates, input to the *Trace-driven Simulator* is the threads constructed by the parallelizing compiler, the traces produced by the execution-driven approach and the parameters from the *Computational Model* that describe the target architecture.

A presentation of the technique used by the parallelizing compiler to construct the threads is beyond the scope of this chapter; for more detail, the interested reader may consult[10, 11]. We now provide a brief overview of the content of the threads and their software implementation.

The threads produced by the parallelizing compiler consist of instructions and synchronization primitives. We use *send* and *receive* type primitives to preserve data dependences in the program and barrier synchronization to preserve control dependences in the program. Table 2.1 illustrates a *Source Program* segment and corresponding threads intended for execution on 2 processors. The source program is written in a high level language such as Pascal, Ada or C. In Table 2.1, statements from the source program have been assigned to either processor 1 (cpu_1) or processor 2 (cpu_2) except for the **while** loop which has been duplicated on both processors. To avoid timing-sensitive problems, we duplicate the conditional branch (*brif*) for the **while** on both processors, but the test of the **while** condition is assigned only to cpu_1. Since the processors execute the threads asynchronously, cpu_1 and cpu_2 proceed to execute statements with a rendezvous occurring only at the barrier.

The semantics of a barrier synchronization are that processors arriving at a barrier must wait at the barrier until all processors arrive. When all processors arrive at the barrier then they may all proceed past the barrier, in this case, to execute the body of the **while** loop. The semantics of the *send* (Sd) and *receive* (Rv) primitives are that the *send* is an asynchronous *send*, so that the sending processor is not required to wait until the receiving processor arrives at the corresponding *receive*. We have implemented both *send/receive* and barrier synchronization in software. For example, for the *send/receive* primitives, we use a bit for each definition in the program. A processor executing a **Sd I** primitive sets the bit corresponding to *I*. The processor executing a **Rv I** primitive checks the bit corresponding to *I*, if the bit has been set then the processor proceeds, otherwise the processor spins in a loop until the bit is set.

2.3. THE COMPUTATIONAL MODEL

To accurately evaluate the partitioning scheme that constructs threads for parallel execution, it is necessary that we be precise about the asynchronous multiprocessor system under study. In particular, we assume that such a system consists of p asynchronous identical processors, *shared global memory modules*, and a *communication structure* that allows processors to communicate with other processors or with the shared memory. An example of such a communication structure is a *bus* that typically allows a single processor to communicate values to memory.

In conjunction with the above system, we use three parameters that, together, describe the "speed" of the architecture. The first is a function $F_e(I)$ that returns the number of cycles required to execute instruction I. The second is a function $F_c = F_a + F_w$, that indicates the number of cycles needed for communication of values through the interconnection structure. For an *interconnection structure* or *communication structure*, we include hardware support such as memory channels[8, 16], register channels[5] or an interconnection network[9] that provides support for communication of values. The function F_a is the access time needed to traverse the communication structure and F_w is the number of cycles a processor waits (due to contention) before it can access a required value. The third parameter, BW, is the *bandwidth* of the communication structure or the number of processors that can simultaneously use the structure. Contention occurs when the number of processors vying to communicate during a given cycle, exceeds BW. The multiprocessor simulator discussed in this chapter takes the parameters p, F_e, F_c, and BW as input.

This computational model is used in both the trace-driven and program-driven simulations. We will discuss the model in more detail in the next section.

2.4. INSTRUCTION-DRIVEN SIMULATION

As discussed in the overview in Section 2.2, we construct threads consisting of statements from the source program and synchronization primitives. We now present the execution-driven multiprocessor simulator which simulates the parallel execution of the threads. Figure 2.3 summarizes the simulator.

2.4.1 The Execution-driven Approach

For the simulator of Figure 2.3, execution begins in Multiprocessor by supplying the parameters F_e, F_c and BW, and p, as discussed in the previous section. Since the simulator must actually execute the statements in a thread, the second statement in the Multiprocessor initializes an *Interpreter* that actually executes the instructions in the thread. After initializing the *Interpreter*, the simulator then reads the parallel *threads*, one thread for each processor. In executing the loop in the main program of the simulator, the CREATE primitive is used to instantiate

```
algorithm    Multiprocessor
input        maxProcessor, the number of processors
local        ref (cpu) array processor [1..maxProcessor]
begin Multiprocessor
    input parameters to specify Fe(I), Fc, BW and p
    initialize Interpreter
    input threads
    for i = 1 to p loop
        processor[i] := CREATE cpu(i)
        ACTIVATE processor[i]
    end for
    HOLD(75000)
    output statistics
end Multiprocessor

algorithm    cpu(i)
input        i, the number of this processor
local        pc, the program counter for this processor
global       busCount, number of processors using the bus
begin cpu
    while there are more instructions in thread_i loop
        case statement_j in thread_i
            send wait to access the bus; increment busCount;
                 set the synchronization bit for this data value;
                 HOLD(cycles required for send); decrement busCount
            receive /* examine a synchronization bit stored in local memory */
                 while synchronization bit for this value is not set loop
                     HOLD(1);
                 end while
                 HOLD(cycles required for receive)
                 reset synchronization bit for this data value;
            barrier wait to access the bus;
                 while number of processors that have arrived at the barrier is
                       less than number of processors using the barrier loop
                     HOLD(1);
                 end while
                 reset the barrier
            store wait to access the bus; increment busCount
                 pass this operation to Interpreter;
                 HOLD(cycles required for store); decrement the busCount;
            load /* similar to store */
            operation pass this operation to Interpreter;
                 HOLD(Fe(operation));
        end case
        increment pc to indicate appropriate instruction in thread_i
    end while
    write('processor', i, 'terminating execution at', time);
end cpu
```

Figure 2.3: Summary of the instruction-level Multiprocessor Simulator.

p processors (*cpu*) and the ACTIVATE primitive is used to begin execution of *thread$_i$* in the respective processor *cpu$_i$*. In *SimCal*, as in Simula, the main program is itself a process and must not be allowed to terminate before any child processes terminate, since that would cause the entire program to terminate prematurely. Thus, the final simulation primitive executed in the main program is HOLD(75000), which inserts the main program at the end of the event list, allowing each of the *cpu$_i$* an opportunity to execute the respective threads. Execution will resume in main after all of the *cpu$_i$*'s have terminated. At that time, any statistics that may have been gathered during the simulation, such as the total time spent waiting to access the bus, may be output.

In addition to the main program of the simulator, a summary of the actions of each processor (*cpu$_i$*) is also illustrated. Each *cpu$_i$* is a *process* that, through the use of the event list, can execute in "parallel". The important part of *cpu$_i$* is a **while** loop that contains a **case** statement that chooses the actions to be performed by the simulator. The important operations listed are: *send, receive, barrier, store, load* and *operation*. By *operation*, we mean an intermediate or assembly code operation.

To illustrate the actions of the multiprocessor simulator, we will now discuss the *send* primitive listed in the **case** statement in process *cpu* shown in Figure 2.3. The first action of the *send* primitive is to "wait to access the bus" as described above. Having gained access, the next action of the *send* primitive is to increment busCount to update the number of processors currently using the bus. Then, the synchronization bit corresponding to the data value being communicated is set to indicate to the receiving process that the value is "ready". Having "sent" the data, the next action in implementing the *send* is to HOLD for the number of cycles that are required in the *send* operation; this execution of the HOLD will update system time appropriately. In the early stages of the multiprocessor simulations, we executed the HOLD primitive for the *send* operation for the number of cycles that we felt were reasonable. Later, as we will discuss in the next section, we conducted experiments on an actual multiprocessor to provide greater accuracy for our simulations/predictions. The final action of the *send* primitive is to decrement the busCount to indicate that this processor is now relinquishing the bus. The actions of the other operations listed in process *cpu* are similar to the *send* primitive.

2.4.2 Validating the Execution-driven Approach

In the previous section we presented the design of the parameterized multiprocessor simulator. In this section, we validate the simulator by using it to simulate the executions of threads produced by our algorithm to partition a sequential source program into threads for parallel execution. This is achieved by supplying appropriate values for the parameters p, $F_e(I)$, F_c, and BW, to the simulator that we construct using $SimCal$[12].

In order to determine the performance of our partitioning scheme on a "real" multiprocessor, we executed our parallel threads on an AT&T 3B2/1000 shared memory multiprocessor system. The 3B2 is equipped with a unibus communication structure, four 32-bit microproprocessors operating at 24 megahertz and an 8 Kbyte physical cache (instruction and data) for each processor. Although there are four processors on the 3B2/1000, one of the processors is dedicated to managing operating system calls. Thus, unless single user access is guaranteed, timing results may be obfuscated for experimentation that uses all four processors. As we will show, we obtained a good correlation between these "actual executions" and the simulations, thereby increasing our confidence in the multiprocessor simulator. The *send* and *receive* primitives were implemented in software on the 3B2 using spin-lock operations on *unix shared variables*. In order to obtain the parameters for our simulator, we first conducted a series of experiments to determine the average cost of the *send* and *receive* primitives and the cost of using the unibus communication structure. These experiments revealed that a *send* primitive requires approximately the same time to execute as a floating point multiplication, and that a *receive* primitive requires approximately twice as long as a floating point multiplication (provided, of course, that the *receive* does not have to wait). These values were utilized in setting the parameter F_e for the simulation studies described below.

Table 2.1 summarizes the results of our experiments using the execution-driven simulator and the 3B2 multiprocessor. The host machine for the simulations was an an IBM PS/2 with an Intel 80486/SX processor; the host machine for the actual executions was the AT&T 3B2. Input programs were both general purpose and scientific in nature. The programs listed in Table 2.1 are: (1) *Trapezoid* program that approximates the area under a curve; (2) *Search* program that searches an array

Comparing simulation & execution					
Test Program	simulations using computational model			actual execution on AT&T multiprocessor		
	cycles p = 1	cycles p = 2	SpUp	time p = 1	time p = 2	SpUp
Trapezoid	8,616	4,393	1.96	7.56	4.02	1.89
Search	43,022	21,533	1.99	15.06	8.58	1.91
Sieve	16,292	11,466	1.42	149.85	106.96	1.40
Vector	64,006	33,999	1.88	26.6	17.34	1.53
Sort	60,342	31,348	1.92	24.03	12.68	1.89

Table 2.1: A comparions of speedups computed using simulations, with speedups computed using executions on an AT&T multiprocessor.

for the largest element; (3) *Sieve*, for computing prime numbers; (4) *vector*, a program by John Rice[14]; (5) *Sort*, Batchers merge-exchange sort[1].

In Table 2.1, the first column lists the programs used in the experiments, the next three columns report the results of the simulations and the last three columns report the results of the actual executions. For the simulations, the second and third columns express the number of cycles required to execute the test program on 1 and 2 processors respectively. For the actual executions, the fifth and sixth columns express the number of seconds required to execute the test program for multiple iterations. Batcher's sort was executed 100 times; search, sieve and vector were exectued 1000 times; trapazoid was exected 10,000 times. These experiments were conducted 1000 times and the results reported are the averages. As a particular instance, note that the simulation indicates that 8,616 cycles are required to execute the sequential code of the Trapezoid program, and that 4,393 cycles are required to execute the schedule for 2 processors with a resulting speed-up of 1.96 over the sequential execution. For the actual execution of the Trapezoid program on the 3B2 multiprocessor, an average of 7.56 seconds were required for 10,000 iterations using 1 processor and 4.02 seconds were required for 10,000 iterations using 2 processors producing a speed-up of 1.89 over the sequential execution.

The similarities in speed-up between the simulation and actual execution results are established by comparing columns 4 and 7. With the exception of the Vector program the difference between these speed-ups is never more than 0.08. The timing results of the Vector program were perturbed by calls to library routines to compute sine and logarithm. This is a remarkably small difference, and certainly validates the use of the simulation approach in most instances.

In addition to supporting the correlation between the simulation results and the actual executions on a 3B2 multiprocessor, Table 2.1 also supports the conclusion that our partitioning scheme is able to provide good speed-up for programs containing sufficient parallelism. Sufficient parallelism implies that the sequence of code being scheduled does not contain a large number of data dependences which can increase communication overhead among the processors.

2.5. COMPARING TRACE-DRIVEN AND INSTRUCTION-DRIVEN SIMULATION

Our approach to trace-driven simulation is similar to that of Johnson[6]. Our instruction traces are generated by the execution-driven simulations (see Figure 2.1) with a trace stream generated by each processor. Thus, in the process-oriented simulations, there will be a thread and a trace stream for each processor (cpu_i was discussed in Section 2.4). In addition to the trace streams and threads, a third input to the trace-driven simulator is the parameters that describe the target multiprocessor. Instruction traces consist of branch target addresses and the addresses referenced by loads and stores. The trace-driven simulator takes the trace stream and parameters as input, and using the threads, generates the full instruction trace needed by the simulator.

Except for the fact that the simulator does not actually execute instructions, its operations closely parallel that of the execution-driven approach. A processor in the simulation fetches an instruction from the thread. If that instruction is a branch, then the simulator consults the dynamic trace stream to determine the target of the branch. If the instruction in the thread is a synchronization primitive, then the simulator must "simulate" the execution of that primitive by setting/testing an appropriate synchonization bit, or by waiting at a barrier. For all instructions, (branches, synchronization primitives and operations), the simulator must account for penalties associated with the execution of these instructions in an actual implementation. This is accomplished by using the HOLD(t) simulation primitive, where t is a parameter from the computational model. Since we model contention in the interconnection network, the simulator must not allow more than BW processors to access the network in parallel. Thus, a processor desiring to access the interconnection network must wait in a fifo queue if BW processors are already accessing the network.

Table 2.2 compares the time and space considerations in the trace-driven and execution-driven approach. The host machine for both approaches was an IBM PS/2 with an Intel 80486/SX processor. The first column of the table lists the programs used in the experiments, the next three columns report the time/space considerations for the execution-driven approach and the final three columns report the time/space considerations for the trace-driven approach. Columns 2 and 5 indicate that the trace-driven approach required from 2 to 19 times as much space as the execution-driven approach. As an example, for the *Search* program, 2,216 bytes of storage were required in the execution-driven approach while 42,224 bytes of storage were required in the trace-driven approach. For the execution-driven approach, only the threads must be stored on disk; however, in the trace-driven approach, both the threads and the traces must be stored on disk. For some applications, very long traces can be generated[2], resulting in greater savings in disk space for the execution-driven approach.

In Table 2.2, columns 3 and 6 indicate that the execution-driven approach required more time to simulate execution of the threads. For example, for the *Search* program, the execution-driven approach required 33.07 seconds to simulate execution of the threads while the trace-driven approach only required 8.84 seconds. Thus, the execution-driven approach required almost 4 times longer to simulate execution of the threads.

Column 4, in Table 2.2, indicates a *confidence* in the ability of the execution-driven simulation to predict the speed-up that might be achieved by actually executing a given thread on an AT&T 3B2 multiprocessor. For example, Table 2.1 illustrates that the *Trapezoid* program achieved a speed-up of 1.96 in the simulation and 1.89 in the actual executions on the 3B2; thus, the ratio of the speed-ups is 1.04 percent. This percentage difference can be used to measure a confidence level in the ability of the simulations to predict performance in an actual execution. Of course the confidence level is the same for both execution-driven and trace-driven approaches (columns 4 and 7) since the threads achieved the same speed-up for a given program in both approaches.

Test Program	Space/Time Comparison					
	Execution-driven approach			Trace-driven approach		
	space	time	confidence	space	time	confidence
Trapezoid	6,440	4.72	1.04	11,484	1.49	1.04
Search	2,216	33.07	1.04	42,224	8.84	1.04
Sieve	23,604	15.87	1.01	40,529	5.99	1.01
Vector	19,499	42.62	22.80	59,058	12.63	22.80
Sort	12,984	28.45	1.50	49,934	21.53	1.50

Table 2.2: This table compares the space and time required in the execution-driven approach with the trace-driven approach.

2.6. CONCLUDING REMARKS

This chapter presents the design of a computational model for a tightly-coupled, asynchronous multiprocessor and the use of that model to construct a parameterized, execution-driven multiprocessor simulator. The multiprocessor simulator is used to execute parallel threads constructed from sequential programs to determine the potential speed-up on various multiprocessor configurations. Through experimentation, we derive parameters for our computational model that describe the AT&T 3B2/1000 multiprocessor and by supplying these parameters to our simulator, we compare the simulations to execution of the threads on the 3B2. The close correlation between the speed-up achieved in the simulations and the speed-up achieved in the actual executions support our conclusion that the simulator can validly predict the performance of a parallelized program on an actual multiprocessor. Finally, we compare the execution-driven approach to the trace-driven approach using traces obtained from the execution-driven simulations. Since trace-driven simulation executes faster than instruction driven simulation, the model that is used as input to the trace-driven simulator can be modified to efficiently test alternative architectural configurations.

REFERENCES

1. K. E. BATCHER. "Sorting networks and their applications", Proceedings AFIPS Spring Joint Computer Conference, pages 307–314, 1968.

2. A. BORG, R. KESSLER, and D. WALL. "Generation and analysis of very long address traces", Proceedings of the 17th International Symposium on Computer Architecture, pages 270–279, May 1990.

3. F. DAHLGREN. "A program-driven simulation model of an mimd multiprocessor", 24th Simulation Symposium, pages 40–49, April 1991.

4. S J. EGGERS and R H. KATZ. "Evaluating the performance of four snooping cache coherence protocols", Proceedings of 16th International Symposium on Computer Architecture, pages 2–15, 1989.

5. R. GUPTA. "Employing register channels for the exploitation of instruction level parallelism" Second ACM SIGPLAN Symposium on Principles and Practice of Parallel Programming, March 1990.

6. M. JOHNSON. Superscalar Microprocessor Design. prentice Hall, 1991.

7. E. KOLDINGER, S. EGGERS, and H. LEVY. "On the validity of trace-driven simulation for multiprocessors", Proceedings of 18th Annual Symposium on Computer Architecture, pages 244–253, May 1991.

8. J. S. KOWALIK. "Parallel MIMD computation: HEP supercomputer & its applications", Scientific Computation Series, 1985.

9. T. LANG. "Interconnections between processors and memory modules using the shuffle-exchange network", IEEE Transactions on Computers, C-25(5), May 1976.

10. B. A. MALLOY, R. GUUPTA and M. L. SOFFA. "A shape matching approach for scheduling fine-grained parallelism", MICRO-25, The 25th Annual International Symposium on Michroarchitecture, pages 131–

135, December 1992.

11. B. A. MALLOY, E. L. LLOYD, and M. L. SOFFA. "Scheduling dags for asynchronous multiprocessor execution", IEEE Transactions on Parallel and Distributed Computing, 5(5):498–508, May 1994.

12. B. A. MALLOY and M. L. SOFFA. "Conversion of simulation processes to pascal constructs", Software - Practice and Experience, 20(2):191–207, February 1990.

13. R. MUKHERJEE and J. BENNETT. "Simulation of parallel computer systems on a shared memory multiprocessor", Proceedings of 23rd Hawaii International Conference on System Science, 1:242–251, 1990.

14. J. R. RICE and J. JING. "Problems to test parallel and vector languages", Technical Report CSD-TR-1016, Purdue University, December 1990.

15. R. RIGHTER and J. C. WALRAND. Distributed simulation of discrete event systems. Proceedings of the IEEE, 77(1):99–113, January 1989.

16. B. J. SMITH. "Architecture and applications of the HEP multiprocessor computer system", SPIE (Real-Time Signal Processing IV), Society of Photo-Optical Instrumentation Engineers, 298:241–248, 1981.

CHAPTER 3

INSTRUMENTATION SYSTEMS FOR PARALLEL TOOLS

Abdul Waheed and Diane T. Rover

3.1. INTRODUCTION AND BACKGROUND

Parallel and distributed computing is becoming popular to achieve a level of performance for a number of applications that is unimaginable with conventional sequential computing. As parallel and distributed computing are helping define new paradigms to fully benefit from the performance gains, the need for tools and technologies to make this type of computing simple and effective for the users can not be understated. Due to the complex nature of concurrent systems and the inherent difficulty in comprehending concurrent activities by human users, development of monitoring tools for parallel and distributed system is a formidable task. These tools are needed to perform diverse tasks, such as program debugging and steering, analysis of bottlenecks in a program, performance prediction, measurement-based performance evaluation, program or performance visualization and animation, etc. There is one common denominator in the complete range of these tools: each relies on runtime information collected from concurrent programs. Unlike data collection from a sequential program, data collection from a parallel or distributed system involves the management of multiple monitoring modules and operating system services. Depending on the nature of an application, this data collection can adversely impact the performance and correctness of the actual program. In order to develop reliable tools, the behavior and overhead of the data collection (monitoring) facility need to be characterized and accounted for.

Software monitoring of parallel and distributed systems usually results in collecting large volumes of data. Visualization is considered a

35

powerful technique for a user to glean useful information from an enormous amount of data. The task of program performance visualization is divided into two steps: the first step is accomplished by *instrumentation* that generates a *trace* of execution of the program on the concurrent system; and the second part is concerned with *visualization* and the representation of this data to depict the behavior of the program during execution[1]. Various visualizations represent different levels of detail and types of information. For example, consider a simple parallel computer system consisting of sixteen processors, connected in a *hypercube* topology. Figure 3.1 depicts four of many possible displays, shown here in the ParaGraph performance visualization tool[2]. The displays provide information about the utilization of the system, the processor interactions (in time and space), etc. Information provided by the utilization summary display is at a higher level in terms of the state of the parallel system, whereas information presented by the space-time diagram shows lower-level interactions among processors. The hypercube display augments the space-time diagram by presenting time via animation and processors topologically. As an alternative to visual representation, the trace data display shows the same information in the form of text. The multiple, distinctive visualizations help the user in performance evaluation and debugging of the program.

The limitations of conventional performance data visualization techniques necessitate better visual analysis and presentation methods that are scalable with the problem and system sizes and extensible. Figure 3.2 represents several examples of advanced visualizations, including higher dimensional displays and use of scientific visualization tools[1].

Visualization and analysis place varying demands on what instrumentation is required and how much data are collected. For instance, the Paradyn tool analyzes performance bottlenecks in a program based on user-supplied thresholds. Its dynamic instrumentation technique defers the insertion of instrumentation code until bottlenecks are indicated[3]. Consequently, it restricts the amount of data collected; however, an additional functionality is required in the data collection modules, so that they can communicate with the application processes during their execution. Program steering is another area where this bidirectional communication is needed between the steering tool and application processes. Steering tools usually present the program behavior to the user through visualization and allow the user to

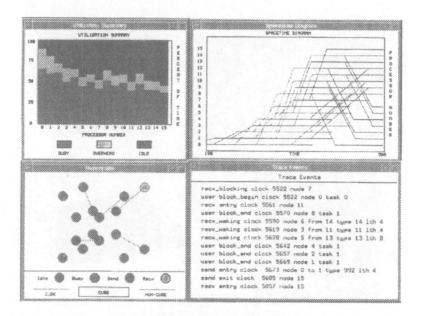

FIGURE 3.1. An example of program performance visualization using a conventional tool (ParaGraph).

interactively change the execution parameters of the application (i.e., steer the application) to enhance its performance[4].

In order to investigate and advance various issues of data collection modules and services used in extant parallel tools, we have introduced the term *instrumentation system* (*IS*). An *IS* consists of modules and functions supported by the parallel or distributed system to collect, manage, and process runtime information obtained from concurrent processes. Due to the expanding range of applications for parallel and distributed systems and the growing number of tool development efforts, the development process of *IS*s needs some scrutiny. This is especially true for applications facing stringent performability criteria, in which excessive or unpredictable perturbation and overheads due to the *IS* can be detrimental or even catastrophic. Therefore, a comprehensive evaluation of the *IS* in early development stages by the tool developers can help avoid problems. Additionally, the properties (flexibility, usability, etc.) of an *IS* can greatly enhance the utility of a parallel tool. Thus, we have studied performance tools from

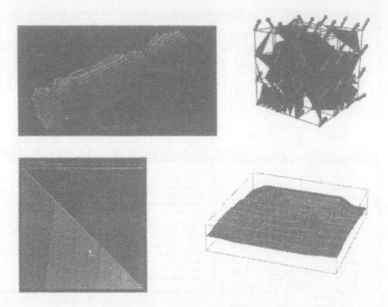

FIGURE 3.2. Examples of advanced performance visualizations.

the perspective of their *IS*s, including *IS* evaluation and development approaches as well as the design of *IS* components and services.

In addition to the critical role of an *IS* for collecting runtime information from a parallel system, it is a key facilitator for setting up an integrated environment of parallel tools[5]. Integrated tool environments may be preferred over individual tools due to the potential added value from extensibility of functions[6]. For example, integration of monitoring tools with modeling and analysis tools has been found useful for performance prediction[7]. Moreover, an integrated environment may provide the infrastructure for interoperability among monitoring tools, debuggers[8], steering tools[4], bottleneck analysis tools[9], correctness-checking tools[10], and visualization tools[2].

This chapter surveys the state-of-the-art in parallel and distributed tools and classifies them according to various features of their *IS*s. Section 3.2 presents a methodology for classifying the *IS*s of various types of parallel tools. Section 3.3 discusses a number of selected tools and characterizes several of their *IS*s on the basis of the methodology presented in Section 3.2. We conclude with a discussion of the significance of this research in Section 3.4.

3.2. A SCHEMA FOR CLASSIFYING *IS*s

This section defines the terminology that we use throughout this paper. Several terms are identical to those used by others in the literature on monitoring systems, which serves to establish some consistency and clarity of discussion in this area. A *monitor* is a tool used to observe the activities of a system. In general, monitors observe the performance of systems, collect performance statistics, analyze the data, and display results. Monitors are used by performance analysts, programmers, system designers, and system managers. For a parallel or distributed system, a monitor is responsible for the collection and analysis of distributed program information[11].

3.2.1. Instrumentation System

An *instrumentation system* (*IS*) is the part of a monitor that is concerned with the collection and management of performance information. The scope of an *IS* spans three modules or functions that are components of any software monitor for a parallel or distributed system. These components are defined in the following subsections. We use the term *instrumentation data* to account for both *execution* information (messages, memory references, I/O calls, etc.) and *program* information (variables, arrays, objects, etc.). We have developed a generic instrumentation system model that represents a majority of components found in extant *IS*s and omits unnecessary implementation details. This generic model is depicted in Figure 3.3. The model defines three components of an *IS* that supports tool integration: (1) local instrumentation servers (*LIS*), (2) instrumentation system manager (*ISM*), and (3) transfer protocol (*TP*). In this paper, we study the *IS*s of selected tools with respect to this model.

The *Local Instrumentation Server (LIS)* captures instrumentation data of interest from the concurrent application processes and forwards the data to other *IS* modules for consumption by appropriate tools. Typically, the *LIS* uses local buffers and a management policy to accomplish data capturing and forwarding functions.

The *LIS* forwards instrumentation data from the concurrent system

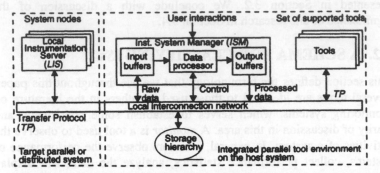

FIGURE 3.3. Components of a typical instrumentation system supporting an integrated tool environment.

nodes to a logically centralized location called the *Instrumentation System Manager* (*ISM*), which manages the data in real-time. The functions of the *ISM* include temporary buffering of data, storing of data on a mass-storage device, and pre-processing of data for analysis and/or visualization tools (e.g., causal ordering).

Instrumentation data are transferred from the *LIS* to the *ISM* and further to various analysis and visualization tools in an integrated tool environment. Usually, a consistent instrumentation data and control *transfer protocol* (*TP*) is used for *IS*-related communications. A majority of existing monitors use operating system-supported interprocess communication abstractions (such as *sockets* in Pablo[12] and Issos[11], *pipes* in Paradyn[9], and *remote procedure calls* in TAM[13]) to accomplish this purpose. Some monitors (such as Hewlett-Packard's VIZIR[8]) implement customized high-level protocols, developed on top of operating system functions, to enhance the flexibility and portability of the instrumentation data transfer and control messaging mechanisms.

3.2.2. Integrated Parallel Tool Environment

A proliferation of parallel program performance analysis, debugging, steering, and visualization tools has led to the development of several integrated parallel tool environments to enhance the usability of individual tools[6]. An *integrated parallel tool environment* supports the use of multiple, possibly heterogeneous tools that cooperate for carrying out one or more analyses of the same parallel program. Tools built by different developers are referred to as *heterogeneous tools* by Hao et al.[8].

An integrated environment may support heterogeneous off-line tool usage, such as TAU[14] and ParaVision[15]; homogeneous on-line tool usage, such as Paradyn[9]; or a combination of the two, such as SPI[10], VIZIR[8], and ParAide[13].

Integrated parallel tool environments rely on particular mechanisms invoked by the *IS* to capture, process, and consume instrumentation data. Figure 3.4 shows the basic technologies in use for tool integration. Tools are integrated with the support of debuggers, operating systems, languages and compilers, or runtime libraries to capture execution and program information from the application processes. Operating system interprocess communication abstractions, such as remote procedure call (RPC), socket, and pipe, are commonly used for transferring instrumentation data. Graphics libraries and graphical user interfaces, such as OpenGL, Tcl/Tk, and X/Motif, provide the user with a consistent view and control of the environment. An *IS* provides a subset of the functionality of an integrated environment, as can be seen by correlating Figure 3.3 with Figure 3.4. Clearly, the *IS* plays a central role in integration; it can cause undesirable and unexpected overhead and perturbation to an application program if its design is not properly evaluated.

3.2.3. *IS* Classification

To address the issues related to its design and evaluation, we classify an *IS* in terms of (1) off-line versus on-line tool usage, i.e., the time constraints imposed by analysis tools in the environment, and (2) *IS* development, management, and evaluation approaches (including any cost models used for evaluation). These dimensions of *IS* classification are defined in this subsection.

3.2.3.1. *Off-line/On-line IS*

An *IS* that supports analysis, debugging, and/or visualization of a parallel application program as a batch process after program execution is called an *off-line IS*. The *LIS* and *ISM* still collect and manage instrumentation data in real-time for such *IS*s. The *ISM* simply stores the data for post-processing.

An *IS* that supports analysis, debugging, steering, and/or visualization of a parallel application program in real-time, concurrent

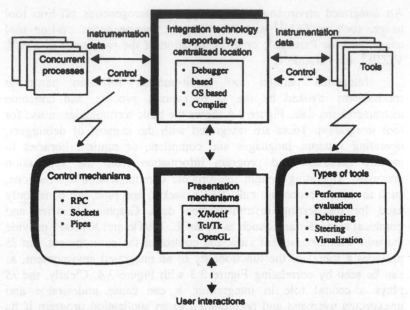

FIGURE 3.4. Basic components and technologies for a typical integrated parallel tool environment.

with application program execution, is called an *on-line IS*. In this case, the *ISM* interacts with the tools and dispatches instrumentation data as soon as on-the-fly preprocessing is finished. When an on-line *IS* supports an integrated tool environment, it maintains a steady flow of runtime data to the tools.

3.2.3.2. IS Development

The process of planning, designing, and synthesizing an *IS* for a particular hardware platform is collectively referred to as *IS development*. The multi-step process is mandated in light of the complexity stemming from system integration issues that need to be addressed. An instrumentation system may be classified as having a *hard-coded* or an *application-specific* implementation, which refers to the approach used to synthesize and integrate the software components in an environment.

Most of the *IS*s found in the literature are designed to complement the analysis and/or visualization functions of a monitoring tool.

Therefore, the primary concern of the design is to provide the user with a self-contained tool. Consequently, the majority of extant *IS*s are "hard coded" into the monitoring tools. The configuration and functions of the *IS* remain unchanged for different application programs, which may have entirely different monitoring requirements. For instance, a parallel program using PICL[16] must initialize instrumentation functions and buffers at all nodes in a multicomputer system, even if only a subset of the nodes is of interest to the user. This incurs undesirable overhead. Similarly, a user can not change the configuration and/or the management policy implemented by the *IS* from one application to another. This further reduces the flexibility of the *IS*.

Development of application-specific *IS*s is a relatively recent phenomenon. Honeywell's Scalable Parallel Instrumentation (SPI[10]) supports customized synthesis of *IS*s. Its *IS* synthesis approach is based on an Event-Action model. *IS* functions are specified by the user as actions taken by the *IS*, in response to the occurrence of specific events. A user specifies the events and actions in terms of an *Experiment Specification Language* (ESL). Therefore, it is possible for the user to specify customized event buffering and instrumentation data forwarding actions that are optimized (with respect to monitoring overheads) for a particular application. Ogle et al.[11] present a similar approach of developing application-specific monitoring functionality in their Issos parallel programming environment. Paradyn supports the *Paradyn Configuration Language* (PCL) for describing its target architecture and operating system and the language-dependent characteristics of the application and platforms[9]. Tuning and Analysis Utilities (TAU[14]) is an integrated performance evaluation environment. Instrumentation is specified by the programmer via a pC++ class library and Sage++ library, and collected data may be analyzed by on-line or off-line tools supported by the TAU environment.

3.2.3.3. IS Management

An *IS* is a combination of the *IS* components (as defined above) and a set of policies that dictate component behavior. *IS management* refers to the policies that are used to schedule various activities of the *LIS* and *ISM* parts of the system. Management of an instrumentation system comprises policies for the data capturing and forwarding functions at the *LIS* level; and for the data collection (from the *LIS*s), on-the-fly processing, storage, and tool interaction functions at the *ISM* level. *IS* management policies streamline the manner in which system resources, such as buffers, CPU time, communication bandwidth, and operating system services, are shared among *IS* and user processes. This subsection classifies the management policies implemented by various monitoring tools in three categories: (1) *static*, (2) *adaptive*, and (3) *application-specific*.

Perhaps the simplest approach to manage an *IS* is to statically specify parameters and policies at the time of development. In this case, *IS* management does not change in response to *IS* load fluctuations (e.g., rate of filling/flushing buffers at various levels of the *IS)* or shared-system resource contention. Instead, policy alternatives are built into the *IS* to remedy known problems or respond to particular conditions. For example, if a buffer becomes full during the execution of an instrumented program, alternatives may include dropping the subsequent logs, overwriting the most recent or least recent logs, and so on. PICL[16]. AIMS[17], ParAide[13], and VIZIR[8] *IS*s employ static management. PICL allows the user to specify the maximum size of an instrumentation data buffer in the code before compilation. If the rate of logging data to a buffer is very high and it becomes full before the program terminates, then by default, no further data are logged on that particular node. The default management policy followed by PICL does not empty buffers automatically prior to program completion. Similar to PICL, AIMS also allows the user to specify buffer size and/or flushing option at compile time. ParAide's TAM stores instrumentation data in *LIS* buffers, which are periodically flushed to an event trace server that forwards them to a tool. VIZIR captures instrumentation data from application processes with the help of debuggers. These data are forwarded on-the-fly to a VIZIR front-end. Although the rate of arrival of this information can vary greatly from one application program to another, the front-end's handling of the data remains the same. Hence, despite simplicity, static management presents the potential for creating new bottlenecks in the system. Adaptive management policies provide one possible solution.

Adaptive management policies differ from static policies in their ability to modify themselves in response to the dynamic *IS* state. For example, an *IS* might modify its data capturing and forwarding policies in response to an excessively high or low frequency of occurrence of certain events. Or suppose the system is constrained due to limited availability of shared resources, e.g., memory space that is shared by user code/data and *IS* event trace buffers. Then the management policy might switch from detailed tracing to sampling to reduce *IS* memory requirements. Two tools that incorporate adaptive management are Pablo and Paradyn. In Pablo, the user specifies one of three possible data capturing mechanisms, listed in decreasing order of perturbation (invasiveness): detailed tracing, event counting, and code profiling. If tracing is active and events with large amounts of associated trace data occur at a frequency higher than that the user desires, the *IS* can dynamically switch to the less-invasive counting mechanism. Paradyn also incorporates support for adaptive *IS* management policies[18]. The overall performance diagnosis approach focuses on locating performance bottlenecks in application programs using the on-the-fly W^3 search model[9]. This establishes a requirement for only sampling some timers and counters. The *LIS* inserts instrumentation into the code only when the analysis tool (part of the *ISM*) needs data to complete its diagnosis[3] and when the cost of the instrumentation is not excessive. The programmer specifies a threshold of maximum tolerable cost for the instrumentation inserted in the program. The *IS* inserts the new instrumentation only when the total cost remains below the specified threshold. If the total cost exceeds the threshold, the instrumentation is deferred and the W^3 search model adapts its strategy for locating bottlenecks. Hence, the adaptive management policy regulates the amount of perturbation. This approach is different from Pablo's approach because the use of dynamic instrumentation ensures the removal of inserted instrumentation when it is disabled[18]. The Pablo *IS* does not remove the instrumentation even if it adaptively switches from detailed tracing to sampling.

Application-specific techniques are tailored to the objectives of monitoring a particular application program. With a customized *IS*, functionality that is unnecessary to meet those objectives can be removed so as to enhance *IS* performance. Customized *IS* management is integrated easily with customized visualization to support specific program performance analysis and visualization. SPI[10], ChaosMON[19], and Issos[11] support application-specific selection of configuration and management policies that minimize *IS* overheads to an application program.

Application-specific management is implementable as either a static or an adaptive approach, possibly under user control depending on an application's monitoring requirements. An adaptive *IS* management approach provides a mechanism that allows the *IS* to be responsive to its load conditions and/or user-specified thresholds to avoid excessive perturbations. However, the development of adaptable analysis and visualization tools is a formidable task, requiring that the tools using instrumentation data modify their operation according to, for example, the volume and frequency of that data.

3.2.3.4. IS Evaluation

An *IS* perturbs the performance of an application program at the *LIS* level due to data capturing and forwarding functions. Additionally, the functions at the *ISM* level can introduce excessive delay between the time that instrumentation data are received by the *ISM* and dispatched to tools. These overheads are the subject of *IS evaluation*. Typically, parallel tools, including *IS*s, are developed in response to users' needs for addressing the performance problems of their application programs. The performance of the tool itself is usually of secondary importance to tool developers. Often, it is the user who discovers a poorly performing tool, for instance, when invoking some feature that unexpectedly and inexplicably causes severe performance degradation.

There are very few examples where tool developers either perform, provide, or document any evaluation of their *IS* overheads through testing with application programs. In particular, we are not aware of this type of evaluation being performed concurrently with the tool design and implementation processes. Paradyn is a notable example in which tool developers provide an adaptive cost model to predict the overhead to an application program due to the *IS*[18]. This cost model is continually updated in response to actual measurements during instrumented program execution. SPI[10] ensures that the invasiveness of its *IS* is accountable. It measures the instrumentation load on nodes and links in each specified window of time to evaluate the degree of invasiveness relative to an application program.

Falcon[4] is perhaps the only tool that supports a thorough evaluation of both *LIS* and *ISM* parts of its instrumentation system. Perturbation to programs is measured at the *LIS* level under different conditions of tracing rates, event record lengths, and event buffer sizes. At the *ISM* level, on-the-fly ordering of event records, which is needed for meaningful visualization, is evaluated as a ratio of out-of-order events that need to be "held back." This *hold-back ratio* is found to be sensitive to the size of *LIS* buffers. Additionally, *IS* performance is compared with other standard instrumentation tools, such as *Gprof*, using the same metrics for overheads. Such meticulous and practical evaluation of *IS* performance by the developers provides essential information to the users, especially when an *IS* is used under real-time constraints.

Presently, it is not standard practice to formally evaluate the performance and functionality of a tool early in its development. Usability and efficiency studies of prototypical tools are emerging to alleviate this situation. However, the underlying *IS* is removed from the end-user and is part of system infrastructure, thus necessitating more rigorous evaluation. Moreover, contemporary approaches to evaluate *IS* overheads and perturbation do not adequately consider the nondeterministic nature of these effects. The approach introduced in this paper has addressed these issues.

3.3. STATE-OF-THE-ART IN PARALLEL TOOLS USING AN *IS*

Preceding sections have identified the significance of an *IS* from two perspectives: (1) as a component of any monitoring-based parallel tool; and (2) as a facilitator to develop integrated tool environments. In order to consider the state-of-the-art in design and evaluation of *IS*s, we classify a subset of the existing tools in terms of their functionality either as a stand-alone tool or an integrated environment. The criteria outlined in Section 3.2 are used to classify these tools. We present an overview of the functions supported by a number of selected tools but specifically focus on characterization and evaluation of the *IS*s of a subset of these tools. This overview is not exhaustive but it is representative of tool design, usage, and evaluation for various applications of parallel and distributed computing.

3.3.1. Overview of Selected Tools

There is a proliferation of tools capable of assisting users of parallel and distributed systems to accomplish various tasks, including performance monitoring and visualization of concurrent programs, analysis of performance bottlenecks, performance modeling and prediction, debugging, and steering. Our objective is to elaborate on the significance of an *IS* (i.e., data collection) to support these tasks and to highlight the functionality of several tools available to perform these tasks. Table 3.1

describes the tasks performed by different types of tools and cites several existing tools as examples in each case.

TABLE 3.1. An overview of various types of parallel tools.

Type of Tool	Description
Performance monitoring	Performance monitoring tools collect instrumentation data from the parallel or distributed system under study. This data is often obtained by inserting instrumentation code in the program or through operating system facilities. Monitoring functions can also be incorporated in system hardware to make it less intrusive. Data obtained from monitoring tools can be used for diverse purposes, such as program performance evaluation, system performance evaluation, or capacity planning and management. Tools such as PICL, AIMS, and ChaosMON support monitoring of parallel systems.
Performance visualization	Performance visualization tools use the instrumentation data collected by the instrumentation system to visually portray the behavior and performance of the program or system. Visualization tools are particularly helpful in analyzing large volumes of data, which are otherwise difficult to comprehend. ParaGraph, AIMS, Pablo, Issos, and ParAide support visualization.
Performance diagnosis	Performance diagnosis tools are used to identify the bottlenecks in parallel and distributed programs. Unlike monitoring and visualization tools that simply present the instrumentation data to the user, diagnosis tools guide the user to locate performance bottlenecks at the source code level. Hollingsworth et al. have incorporated a W^3 search algorithm in Paradyn tool to identify performance bottlenecks[9]. Poirot is another example of diagnosis tools[6]. Such tools also rely on the data obtained from the IS.
Performance modeling and prediction	Performance modeling and prediction tools use the instrumentation data collected by an IS to parameterize a statistical model of the program behavior. A model is often parameterized through a scaled-down execution of the actual program. This model is subsequently used to predict various performance metrics related to the performance of the program. AIMS is one tool that automates the modeling and prediction tasks for concurrent programs [17]. Crovella et al.[7] also describe prediction tools that follow a similar approach.
Program debugging	Program debugging tools help the programmer to locate bugs in their programs at source code level. However, this task is considerably more difficult than debugging sequential programs. The complexity of this task arises from the concurrency of multiple processes that interact with one another. Despite a considerable research effort in this area, the development of a debugging tool for users of a concurrent system is still a non-trivial problem[6]. ParAide and VIZIR are examples of existing parallel debuggers.
Program steering	Program steering tools allow programmers to dynamically alter the behavior of their programs. In order to accomplish this goal, the IS collects the necessary information about the performance and behavior of the system, which may be presented to the human user through visualization. Based on this information, the user interacts with the IS to trigger a dynamic change in the operating parameters of the program. Falcon is an example of a steering tool.

Table 3.2 identifies a number of existing tools and their functions. Some of these tools, such as Pablo, Paradyn, Falcon, ParAide, Prism, SPI, and VIZIR, support integrated environments. In these cases, an IS plays an important role to accomplish the integration, in addition to collecting runtime information.

TABLE 3.2. Classification of selected tools in terms of their functionality.

Tool	Platform	Functions	Reference
PICL	Distributed-memory parallel systems	Performance monitoring (tracing)	http://www.netlib.org/picl/
AIMS	Distributed-memory parallel systems and cluster of workstations	Performance monitoring, modeling, simulation, and prediction	http://fi-www.arc.nasa.gov/fic/parallel/aims.html
Pablo	Distributed-memory parallel systems	Performance evaluation and visualization	http://bugle.cs.uiuc.edu/
Paradyn	Distributed-memory parallel systems and cluster of workstations	Performance evaluation	http://www.cs.wisc.edu/~paradyn/
Falcon/Issos/ ChaosMON	Distributed- and shared-memory parallel systems and cluster of workstations	Performance evaluation, steering, and visualization	http://www.cc.gatech.edu/systems/projects/FALCON/index.html
SHRIMP monitoring software	Research prototype architecture (SHRIMP)	SHRIMP software includes performance monitoring utility	http://www.CS.Princeton.EDU/shrimp/
TAU	Parallel systems using pC++	Performance tuning and analysis	http://www.cs.uoregon.edu/paracomp/tau/
ParaVision	Parallel systems	Integrated environment of performance evaluation, debugging, and visualization	[15]
ParAide	Distributed-memory parallel systems	Integrated environment of performance evaluation, debugging, and visualization	http://www.ssd.intel.com/paragon.html
Prism	Distributed-memory parallel systems	Integrated environment of performance evaluation, debugging, and data visualization	http://cmns-sparc.think.com:80/tmhtml/ProdServ/Products/prism.html
SPI	Distributed-memory parallel systems and cluster of workstations	Performance evaluation and debugging	[10]
VIZIR	Cluster of workstations	Performance visualization and debugging	[8]
PV	Distributed- and shared-memory parallel systems and cluster of workstations	Performance analysis/tuning and debugging	http://www.almaden.ibm.com/watson/pv/pv.html

3.3.2. *IS* Characterization of Selected Tools

In this subsection, we classify the *IS*s of several representative parallel tools, which are a subset of the tools listed in Table 3.2. These are summarized in Table 3.3 according to the following classification features as outlined in Section 3.2: *off-line* or *on-line* performance analysis and visualization; nature of the *LIS* and *ISM* components; *hard-coded* or *application-specific* development of instrumentation software; *static, adaptive,* or *application-specific* management of instrumentation data; and any integral evaluation techniques. The reason for selecting these particular tools is the availability of adequate information regarding their *IS*s through published literature and on-line information.

TABLE 3.3. Summary of *IS* features of some representative parallel tools.

Tool	Analysis /Vis. Support	LIS	ISM	Synthesis Approach	Mgmt. Approach	Eval. Approach
PICL	Off-line	Local buffers using runtime library	Trace file	Hard-coded	Static	—
AIMS	Off-line	Library	Trace file	Hard-coded	Static	—
Pablo	Off-line	Library	Trace file	Hard-coded	Adaptive	—
Paradyn	On-line	Local daemon	Main Paradyn process	Application-specific by using PCL	Adaptive	Adaptive cost model
Falcon/ Issos/ ChaosMON	On-/Off-line	Resident monitor	Central monitor	Application-specific	Application-specific	Evaluation of the factors that affect perturbation
ParAide (TAM)	On-/Off-line	Library	Event trace server	Hard-coded	Static	Accountable invasiveness
SPI	On-/Off-line	Library	Event-Action machines	Application-specific	Application-specific	Accountable invasiveness
VIZIR	On-/Off-line	Library	VIZIR front-end	Hard-coded	Static	—

3.3.3. Evaluation of Selected *IS*s

In order to develop an *IS* in a structured manner, we proposed a rapid prototyping, two-level approach, as depicted in Figure 3.5[20]. On a higher-level, requirements of the *IS* are either determined by the developer or specified by the tool users. These requirements are transformed to detailed lower-level system specifications, which are subsequently mapped to a model representing the structure and dynamics of the *IS*. This model is parameterized and evaluated with respect to chosen performance metrics that reflect the critical *IS* overheads to the application program as well as the target system. The evaluation results are then translated back to the higher-level, so that conclusions can be drawn by tool developers and users regarding *IS* performance. Feedback from the *IS* prototyping process is used to modify either the requirements or the system specifications to obtain desired performance. Finally, the model becomes the blueprint for actual synthesis of the *IS*. More specifically, we are applying object modeling techniques in this process with the intent of using object-oriented software engineering methods to translate the abstract system model into the software modules for the actual system[21].

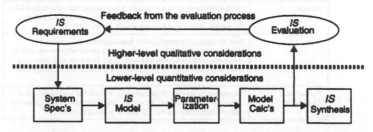

FIGURE 3.5. Two levels of a structured *IS* development approach.

Realization of a tool in general, and an *IS* in particular, is a non-trivial process requiring many person-hours of programming effort. Moreover, evaluation of a tool by users upon its release typically leads to requests for corrections, changes, or enhancements in its function. In contrast, rapid prototyping and preliminary evaluation of an *IS* can be applied to ensure that specific requirements of a tool environment are met prior to the investment in programming effort. This process is likely to deliver better performance, be less costly, and yield greater user satisfaction.

Performance evaluation studies are currently being pursued on developing appropriate evaluation techniques for *IS*s[20]. We have applied this approach to an existing *IS* (PICL)[16] and are currently applying it to two tools: Paradyn *IS* and Vista *IS*[20]. A good example of the use of this approach by others is the measurement-based evaluation of the Falcon *IS* by the developers[4]. Key results of these efforts are summarized in Table 3.4.

Evaluation of an *IS* requires a substantial amount of low-level information about the design of the *IS* from tool developers. However, such a study during the early stages of tool development can lead to a better design of the *IS*. This is a worthwhile effort because tools will only be successful if a proper framework (i.e., *IS*) exits to support them.

TABLE 3.4. Summary of key results of evaluating selected *IS*s.

Tool	Key Evaluation Results
PICL	Management policy for the local trace data buffers during the execution of an instrumented program is critical to determine the *IS* overheads to the program [20].
Paradyn	Number of processes on a workstation that generate instrumentation data is a key factor in determining the overhead to the program as well as throughput of the Local Instrumentation Server (*LIS*) [20].
Vista	Rate of arrival of instrumentation data samples or the number of processes that produce these samples affects the algorithm being used by the Instrumentation System Manager (*ISM*). However, data processing delay at the *ISM* is not very sensitive to differences in the configuration of the input buffers [20].
Falcon/Issos/ChaosMON	Depending on the type of platform (e.g., a cluster of workstations or a massively parallel system), there is a tradeoff between putting extensive analysis functionality at the *LIS* level or at the *ISM* level [4,19,11].

3.4. DISCUSSION AND CONCLUSIONS

The functional requirements imposed on performance tools by today's parallel and distributed systems and applications warrant the use of sound design practices in tool development. However, it is generally realized that tool development is a maturing area; users must demand better tools and developers must deliver them. Fortunately, the software technology necessary to develop tools is becoming more sophisticated and accessible. What is needed is the framework for applying these technologies in an appropriate manner to meet tool specifications and thus user needs. Such a framework supports structured development approaches, tool integration, and a common context for tool evaluation.

In this chapter, an instrumentation system is defined as the framework. We have motivated the development of an *IS* based on both the need to serve and manage a proliferation of diverse performance tools (from various developers and for various users) and the fact that an *IS* exists in some form for most tools. Thus, developers and users alike will benefit from a more coordinated effort in tool development and deployment, which is driven naturally by a common understanding of the functions and structure of an *IS*. Through the discussion in this chapter, we have presented an overview of the current state-of-the-art in *IS*s, relative to a number of tools being served. Moreover, many of the issues critical to developing *IS*s that effectively support integrated tool environments and perform within specific constraints are addressed. These include data management policies, identification of key metrics to evaluate *IS* performance, evaluation methods, and *IS* modeling and rapid prototyping. With progress in these areas of *IS* development, we can move closer to the goals of meeting user needs with better tool environments and providing developers with a foundation for building usable, mature tools.

REFERENCES

1. M. HEATH, A. MALONY, and D. ROVER, "The Visual Display of Parallel Performance Data," IEEE Computer, 28, 11, (November 1995).

2. MICHAEL T. HEATH, and JENNIFER A. ETHERIDGE, "Visualizing the Performance of Parallel Programs," IEEE Software, 8, 5, (September 1991), pp. 29–39.

3. J. K. HOLLINGSWORTH AND B. P. MILLER, "Dynamic Control of Performance Monitoring on Large Scale Parallel Systems," Proc. of Int. Con. on Supercomputing, (Tokyo, Japan, July 19–23, 1993).

4. WEIMING GU et al., "Falcon: On-line Monitoring and Steering of Large-Scale Parallel Programs," Tech. Report GIT–CC–94–21, (1994).

5. DIANE T. ROVER, "Performance Evaluation: Integrating Techniques and Tools into Environments and Frameworks," Roundtable, Supercomputing '94, (Washington DC, November 14–18, 1994).

6. Workshop on Debugging and Performance Tuning of Parallel Computing Systems, (Chatham, Mass., Oct. 3-5, 1994).

7. MARK E. CROVELLA AND THOMAS J. LEBLANC, "Parallel Performance Prediction Using Lost Cycles Analysis," Proceedings of Supercomputing '94, (Washington, DC, Nov. 14–18, 1994).

8. MING C. HAO, ALAN H. KARP, ABDUL WAHEED, and MEHDI JAZAYERI, "VIZIR: An Integrated Environment for Distributed Program Visualization," Proc. of Int. Workshop on Modeling, Analysis and Simulation of Computer and Telecommunication Systems (MASCOTS '95) Tools Fair, (Durham, North Carolina, Jan. 1995).

9. BARTON P. MILLER et al., "The Paradyn Parallel Performance Measurement Tool," IEEE Computer, 28, 11, (November 1995).

10. DEVESH BHATT et al., "SPI: An Instrumentation Development Environment for Parallel/Distributed Systems," Proc. of Int. Parallel Processing Symposium, (April 1995).

11. DAVID M. OGLE, KARSTEN SCHWAN, and RICHARD SNODGRASS, "Application-Dependent Dynamic Monitoring of Distributed and Parallel Systems," IEEE Transactions on Parallel and Distributed Systems, 4, 7, (July 1993), pp. 762–778.

12.   DANIEL A. REED et al., "The Pablo Performance Analysis Environment," Technical Report, (Dept. of Comp. Sci., Univ. of Ill., 1992).

13.   BERNHARD RIES, R. ANDERSON, D. BREAZEAL, K. CALLAGHAN, E. RICHARDS, and W. SMITH, "The Paragon Performance Monitoring Environment," Proceedings of Supercomputing '93, (Portland, Oregon, Nov. 15–19, 1993).

14.   D. BROWN, S. HACKSTADT, A. MALONY, and B. MOHR, "Program Analysis Environments for Parallel Language Systems: The TAU Environment," Proc. of the Second Workshop on Environments and Tools For Par. Sci. Computing, (Townsend, Tennessee, May 1994), pp. 162–171.

15.   GARY J. NUTT and ADAM J. GRIFF, "Extensible Parallel Program Performance Visualization," Proc. of MASCOTS '95, (Durham, North Carolina, Jan. 1995).

16.   G. GEIST, M. HEATH, B. PEYTON, and P. WORLEY, "A User's Guide to PICL", Technical Report ORNL/TM-11616, (Oak Ridge National Laboratory, March 1991).

17.   JERRY YAN, "Performance Tuning with AIMS—An Automated Instrumentation and Monitoring System for Multicomputers," Proc. of the Twenty-Seventh Hawaii Int. Conf. on System Sciences, (Hawaii, January 1994).

18.   J. K. HOLLINGSWORTH and B. P. MILLER, "An Adaptive Cost Model for Parallel Program Instrumentation," Technical Report, (Oct. 1994).

19.   CAROL KILPATRICK and KARSTEN SCHWAN, "ChaosMON—Application-Specific Monitoring and Display of Performance Information for Parallel and Distributed Systems," Proc. of the ACM/ONR Workshop on Parallel and Distributed Debugging, (Santa Cruz, California, May 20–21, 1991).

20.   A. WAHEED and DIANE T. ROVER, "A Structured Approach to Instrumentation System Development and Evaluation," Proceedings of Supercomputing '95, (San Diego, California, Dec. 4–8, 1995).

21.   A. WAHEED and DIANE T. ROVER, "Vista: A Framework for Instrumentation System Design for Multidisciplinary Applications," to Appear in the Proc. of MASCOTS '96, Tools Fair, (San Jose, Feb. 1–3, 1996).

CHAPTER 4

A METHODOLOGY FOR THE PERFORMANCE MODELLING OF DISTRIBUTED CACHE COHERENT MULTIPROCESSORS

A.J. Field and P.G. Harrison

4.1. INTRODUCTION AND BACKGROUND

In this chapter we present a general method for developing of performance models of shared-memory computer systems. These machines comprise a number of architecturally identical processing nodes whose activities are coordinated via a single global shared memory system. Each node has the ability to cache a part of the shared memory in separate high-speed stores local to each processor and the role of the coherency protocol is to ensure that each processor's view of the shared memory is the same at every instant.

In a *distributed* shared-memory system cache coherency is maintained by the interchange of messages between nodes, thereby introducing an overhead on selected read and write operations from the processor. Performance models are vital in order to understand the behaviour of speculative designs and to perform a quantitative comparison of different design options, in particular to measure the effect of the overheads of various cache coherency mechanisms which may be of interest to the designer. A change to the architecture of a processing node, for example, may affect the flow of coherency traffic and create an unexpected bottleneck. Equally, a change to the cache line states

55

may change the amount or nature of the memory and network traffic created for a given workload. Experiments with other hardware parameters such as cache line size are also important for optimizing static hardware parameters and cost/performance.

A commonly used modelling technique in this context is *execution-driven* simulation in which a parallel program is executed on top of a simulated memory system. These simulations can be extremely accurate but incur very long execution times and are highly prone to logical errors in coding. An execution-driven model of SCI has been developed at the University of Edinburgh[1] but this does not explicitly model the communication network.

The alternative, yet complementary, approach proposed here is that of analytical modelling using a combination of established mathematical results and numerical techniques. These analytical models are in general far more efficient computationally than the equivalent simulation. The execution time of the models we describe here are measured in seconds or minutes using Mathematica[2] on a Macintosh computer, compared with many hours, or even days, for a simulation run.

A number of such models have been developed for shared-memory systems. Modelling of bus-based multiprocessor systems has been addessed in some detail[3] and numerical predictions have been produced for a range of operating parameters and coherency protocols. A simple model of a distributed shared-memory system based on the SCI[4,5] protocol has been developed[6], but this only models the ring (the default network for SCI systems) and does not take into account the all-important coherency traffic produced by the SCI protocol itself.

We concentrate here on distributed cache coherency protocols and show how they can be modelled using a general-purpose analytical approach which can be adapted to different architectures and coherency mechanisms. The methodology focuses in particular on the coherency protocol since a specific node architecture can be modelled using standard techniques, e.g. queueing networks.

We demonstrate the approach with two case studies, the first being a model of SCI referred to above and the second a protocol, which refer to as ALITE[7], which bears many similarities also to Sun's S3MP protocol[8]. For each of these we show how the methodology is applied to yield a model, and show some sample results demonstrating the usefulness of the model for experimental purposes. We include some discussion of the queueing models for the node and network although this is not emphasised in any detail. By way of variety, the first model includes a detailed representation of the node architecture and associated bus traffic, but a very straightforward model of the network. The second conversely employs a more substantial network model and a simplistic queueing model of the node.

The rest of the chapter is organised as follows. Section 4.2 describes the class of machines which are the subject of the work. Section 4.3 describes the modelling methodology. Sections 4.4 and 4.5 describe two case studies showing how the methodology can be applied to systems of genuine practical interest. Some numerical results are presented in Section 4.6 and the conclusions are laid out in Section 4.7.

4.2. SHARED MEMORY SYSTEMS

The simplest shared-memory systems consist of a collection of processing nodes together with a global memory all attached to a shared bus. Each processor has a local cache which stores recently accessed locations in the global address space. A *coherency protocol* ensures that the cache contents are consistent, so that each read from a processor always picks up the latest value written. This protocol exploits the broadcast nature of the shared bus which allows each cache controller to listen to all memory transactions from the other processors (*snooping*).

Bus-based machines are conceptually simple but do not scale well so much attention is now being focused on *distributed* shared memory implementations in which the memory is partitioned into a number of segments, one per processing node. The nodes are then connected by a more general, and scalable, interconnection network. The global architecture of such a system is shown in Figure 4.1. Each node now consists of a processing resource (this may be a single processor or a collection of tightly coupled processors, possibly with one or more local caches), a node cache containing copies of the remote locations that are cached at this node, a segment of the global memory, and a controller for managing network communication and remote coherency traffic. Coherency of the caches is maintained by a series of explicit point-to-point communications (c.f. snooping on a shared bus).

In order to exploit spatial locality the memories and the caches are divided into *lines* which are collections of adjacent memory locations. Lines carry with them additional *state* information which is used by the coherency protocol to determine which operations need to be

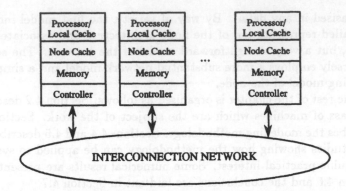

Figure 4.1: Global System Configuration

undertaken in response to each read or write to the line in order to maintain coherency. These coherency operations may change the state of one or more of the cache lines.

Data is shipped around the machine in units of lines and the protocol is usually carefully constructed so that whole lines (which may be of the order of tens or hundreds of bytes) are only transmitted between nodes when absolutely necessary. Where possible, shorter "control" messages are used.

To maintain coherency, it is neccesary to know exactly which node caches in the system contain copies of a given line in memory. This information may be contained in a table, or *directory* at the home node, or it may be distributed in the form of a *sharing list* which links (lines in) the node caches that contain copies of the cached line. Sharing lists may be either singly- or douly-linked.

4.3. MODELLING METHODOLOGY

The proposed methodology breaks down the modelling process into six steps:

1. Definition of the machine configuration and characterisation of the system workload

2. Development of a (finite state space, irreducible and positive recurrent) Markov model of the sharing list states to determine the

equilibrium probability distribution of the number of sharers of a given memory line

3. Determination of the cache line states and state transitions together with the solution of an associated Markov model of the line states to yield equilibrium line state probabilities

4. Determination of coherency operations and the associated probabilities for each in a given state for each type of memory access

5. Identification of the major traffic classes and the number of each incurred by each coherency operation in each state

6. Solution of queueing model(s), or similar, for the communication network and/or node using the traffic rates derived from the previous step

As we are focusing on the modelling of the coherency protocol, we only detail steps 1—5. There are well established techniques for modelling a variety of communication networks and node architectures so we present few specific details. Two examples are, however, briefly discussed as part of the case studies described later.

The performance measures of interest are numerous and are easily extracted from the resulting models; in what follows we focus on the probability that a processor is busy doing useful work, π and the average memory response time (or latency), R.

We now expand the details of the main steps in the methodology.

4.3.1. Preliminaries

It is assumed that each processor alternates *think periods* and periods when it waits for a memory access to be serviced. The mean think period is defined throughout to be $1/\tau_0$. After a think period, the processor generates a memory request and this request may or may not invoke a coherency operation. During the time a request is processed, the processor is idle. Thus, $\pi = \tau_0^{-1}/(\tau_0^{-1} + R) = 1/(1 + \tau_0 R)$. We will write $\tau = \pi\tau_0$ for the rate at which a processor actually generates a read/write request to the memory system.

The machine configuration parameters must be specified and in what follows we assume the following notation:

K - The number of nodes

N – The total number of shareable memory lines

n – The capacity (in lines) of each local cache

The degree of locality in the workload is characterised by its cache hit/miss rates on reads/writes, and by a parameter P_{loc} which is the probability that a memory reference from a processor is to a line whose home is in the local memory of that processor's node. The read hit and read miss rates will be written β_{rh} and β_{rm} respectively; β_{wh} and β_{wm} similarly for writes. It is also convenient to define $\beta_h = \beta_{rh} + \beta_{wh}, \beta_m = \beta_{rm} + \beta_{wm}, \beta_r = \beta_{rh} + \beta_{rm}, \beta_w = \beta_{wh} + \beta_{wm}$.

4.3.2. Sharers

In general, the sharers of a memory line are represented explicitly by a sharing list or similar structure. To write down the transition rates between cache line states (see below), and also to determine the average traffic generated per memory reference, we need information about the number of sharers of a given cache line. This is produced from a separate model of line sharing, taken from the point of view of the memory.

We assume that the evolution of the number of sharers of a memory line follows a Markov process, independent of the states of other memory lines. This process is irreducible, aperiodic and has a finite state space. It thus has a steady-state. The model can be solved using standard techniques since the transition rates are expressed solely in terms of known model parameters. It leads to the mean Δ, and probability distribution of the number of sharers at equilibrium.

4.3.3. Line States

The specification of a cache coherency protocol includes a definition of the possible cache line states and valid state transitions. In the first part of the modelling process we aim to determine the equilibrium probability that a cache line is in each of the defined states.

We again assume that the evolution of the state of a given line in a cache (we shall refer to this as the *observed* cache line) follows a Markov process, independent of the states of other lines.

Unlike the submodel of line sharing, however, the transition rates here may in general be defined in terms of other variables which are themselves dependent on the equilibrium line state probabilities. For example, in a protocol which distinguishes clean data from dirty data a transition rate may depend on the probability that a given cache line (other than the observed one) is in the dirty/clean state. This probability is itself estimated from the equilibrium line state probabilities. This leads to a circularity in the definition of the transition rates and hence to a fixed point equation which is solved iteratively. We shall see examples of this circularity in both the case studies which follow.

If S is the set of line states then the equilibrium line state probabilities will be written $q_s, s \in S$.

4.3.4. Coherency Operations

The next step is to define the coherency operations for the protocol in question. These consist of the various actions required to maintain the identifiers of all the sharers of each cached line. Where explicit sharing lists are used these will include operations for creating new lists, deleting from and adding to existing ones and for reducing or updating lists, depending on whether the protocol is invalidation based or update based.

Some operations will only be valid for certain line states. For example, a reduction operation following a write hit may only take place when the line is cached in at least one other node. This information may be built into the line state, e.g. by a bit distinguishing shared/exclusive status.

In order to compute the mean traffic per memory reference we need to determine the probability that a given operation will take place in a given state. This will be zero when the operation is undefined in a particular state.

We will use Ω to denote the set of coherency operations. The probability that operation $a \in \Omega$ is performed when the observed line is in state $s \in S$ will be written $\delta_{s,a}$.

4.3.5. Node and Message Traffic

The architecture-specific models of the processing nodes and communication network require inputs representing the arrival rates of the various traffic types assumed in the model. For example, in the case of the network, there may be a number of message types used by the protocol, each imposing its own demand on the network resources. Similarly for the nodes, there may be a number of internal transaction types, e.g. bus, memory or cache requests, or even combinations thereof.

There are thus a number of "traffic" types of interest and the next step is to list the number of instances of each type for each state and coherency operation. This information is presented as a set of tables (or a single three dimensional table), one for each transaction type. If there are z transaction types labelled $1, 2, ..., z$ then table $Z_{s,a}^{(i)}$ will denote the number of instances of transaction type i when performing operation a in state s, $1 \leq i \leq z, s \in S, a \in \Omega$.

In order to convert this information into a set of traffic rates for a queueing model of the node or network we need also to enumerate the costs of each transaction. A second table therefore lists the time T_i associated with transaction type $i, 1 \leq i \leq z$.

Sepcifying these tables is not easy; they typically contain a lot of information at a very low level. For notational brevity, in the two case studies below the various transaction tables are coerced into a single table within which the $Z_{s,a}^{(i)}$ and T_i are implictly defined. Reconstructing the individual tables is, however, a straightforward programming task.

4.3.6. Node and Network Modelling

The final part of the process entails modelling the internal node architecture and communication network. These details are, however, decoupled from the protocol. The protocol does not extend to a definition of the processing engine(s) at each node, or to the architecture of the communication network. (Note that the SCI protocol defines a "default" network architecture, namely a ring, but this is not imposed by the standard.)

The details of the node and network submodels are therefore beyond the scope of the methodology. The techniques which can be used are, however, well established. A number of standard text books[9,10] provide some excellent examples.

4.4. CASE STUDY—ALITE

The ALITE protocol is representative of a number of protocols based on unidirectionally linked sharing lists. It is not the optimal protocol but serves as a good reference model. It is similar in many respects to

Sun's S3MP protocol[8] and the Stanford DASH protocol[11] although no direct comparisons should be made.

Although we emphasise the model of the coherency protocol, it is instructive to see how the model interacts with the queueing model of the node. This example includes a very detailed model of the node, in particular with respect to the various types of bus traffic generated by the protocol. This in turn requires the protocol model to produce traffic rates for each of these classes. We explain the latter in some detail but omit many of the details of the node model itself. The network is assumed to be contention free so that the message transmission time is proportional to the message length.

The protocol is invalidation based. On a write, this entails sending an invalidate request to the addressed cache line's home node which marks the home copy as being invalid and then forwards the request down the sharing list. With the exception of the writing node all sharing list entries will be invalidated by setting a bit in the associated line in the second-level cache. The last entry on the sharing list responds to the invalidate request by sending a completion message to the requesting node. In the event of the write being a miss this message will also carry a copy of the line's (previous) data.

When a write is complete the locally cached copy is marked as being *dirty*, i.e. inconsistent with the original home copy, and *exclusive*, meaning that it is the only valid copy. So long as the line remains in the exclusive state the processor may write to it arbitrarily without further communication.

If another processor tries to read a line that has been written by a remote processor the read request sent to the home node is forwarded to the first entry of the sharing list. This will subsequently supply a (valid) copy of the line to the reader and the sharing list links will be updated to include the new sharer at the head. At this point both copies of the line are marked as dirty and *shared*. A line can thus be read and written remotely without having to update the home copy of the line; all that is required is to maintain the sharing list. Only when the final cached copy of a line is displaced from the cache is it necessary to write the line back to memory.

If a second-level cache line forming part of a sharing list is displaced following a miss on another location which maps to the same line, the cached copy has to be "unhooked" from the list. This is achieved by sending a message to the home node which is passed down the sharing list until it reaches the entry preceding the unhooking node. The identifier of this node is then passed to the unhooker which decouples itself from the list by replying with the identifier of its successor. This is used to update the list pointers in the obvious way. We assume that the pointers associated with the second-level cache lines are stored in the network controller so that pointer maintenance and traversal can be performed without generating internal bus traffic.

Note that a line may be copied into the cache at the home node. In this situation the line state is maintained as for any other cached copy except that it does not explicitly appear in the sharing list for that line. Status information at the home memory indicates whether a copy of the line is held in the local cache.

4.4.1. Line States

The second-level cache line states are therefore as follows:

1. **Home Exclusive Clean**—The line is cached at the same node as the home copy. It is the only copy and it has not been written since being cached.

2. **Home Exclusive Dirty**—As above, except that the line has been written and so is inconsistent with the home copy.

3. **Home Shared Clean**—The line is cached at the same node as the home copy and there is at least one other copy cached at another node. The cached copy is clean.

4. **Home Shared Dirty**—As above except the cached copy is dirty with respect to the home copy.

5. **Client Exclusive Clean**—The home copy of the cached line is on another node. This is the only cached copy, however, and it has not been written since being cached.

6. **Client Exclusive Dirty**—As above except that the line has been dirtied.

7. **Client Shared Clean**—Same as 3 except that the line is not at the home node.

8. **Client Shared Dirty**—Same as 4 except that the line is not at the home node.

9. **Invalid**—The line contains no usable information.

4.4.2. Coherency Operations

The coherency operations are listed below. Note that two message types are distinguished: short messages are those which contain only control information, e.g. for managing updates to a sharing list, and long messages are those which contain both control information and a line of data. Long messages are typically about an order of magnitude longer than short messages.

Creation (CR) A read or write miss on a line in state 9 not cached by other processors. A new list is created unless the read/write is initiated from the home node.

Addition (AD) A read miss on a line in state 9 but cached by other processors. The new sharer is added to the existing sharing list.

Reduction (RE) A write hit on a line in state 3—8. The sharing list of which the write is a part is reduced so that the write becomes the only element.

Deletion-Creation (DC) A read or write miss on an uncached line whose address maps to a line in state 1—8 in the cache. The reader/writer is unhooked from the list of which it is a part. A new list is created as per CR above.

Deletion-Addition (DA) A read miss on an already cached line whose address maps to a line in state 1—8 in the cache. This must first be displaced from the cache as above. The node is added to the new sharing list as in AD above.

Deletion-Reduction (DR) A write miss on an already cached line whose address maps to a line in state 1—8 in the cache. This is first displaced and the list associated with the line being written is invalidated. The reduction is similar to RE above except that last item in the sharing list returns a long message to the requester containing a copy of the line.

Invalid-Reduction (IR) A write miss on a line in state 9 cached by other processors. Reduction proceeds as per DR above.

Note that since sharing lists are singly linked, each deletion operation (DA, DC, DR) will require an unhook message to traverse half the mean length of a sharing list on average. Note also that reduction operations (DR, HR, IR) here do not require the writer to take the head of the list before broadcasting an invalidate message to all sharers.

4.4.3. Sharers

The Markov process state transition diagram for the number of sharers is shown in Figure 4.2. Note that the state 0 covers both the case where there is no cached copy of the line and the case where the only cached copy is at the home node. This state can be entered from state 1 as the result of a displacement at the (only) remote node with a copy of the line, and from any other state as the result of a write to the line from the home node. The former occurs with probability P_{miss}/n and the latter with probability $\beta_w P_{loc}/(N/K)$ where $P_{miss} = \beta_{rm} + \beta_{wm}$. The state 1 can be reached from state 0 as the result of a remote read to the line (probability $(K-1)\beta_r(1 - P_{loc})/(N/K)$), from state 2 as the result of a displacement (probability $2P_{miss}/n$), and also from any other state as a result of a write to the line from any non-home node (probability $(K-1)\beta_w P_{rem}/(N(K-1)/K)$ where $P_{rem} = 1 - P_{loc}$). The transition probabilities for the general case are shown in the diagram.

The model is solved to obtain the mean length of the sharing list, Δ, and the equilibrium probability P_i that the sharing list is of length $i, 0 \le i \le K - 1$.

4.4.4. Transition Rates

The transition rates which follow have all been divided by the factor τ which is the rate at which the processor leaves the think state. η_c and η_d are the probabilities that a line is cached in the clean/dirty states respectively and P_c and P_u are the probabilities that a line is cached/uncached respectively. These are estimated straightfowardly from the steady-state probabilites (hence the fixed point equation) and the above sharer model.

$$s \to 1, s \ne 3 \qquad \frac{\beta_{rm} P_u P_{loc}}{n}$$

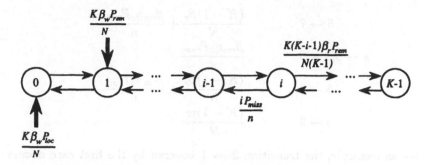

Figure 4.2: Markov model of the number of ALITE line sharers

$$3 \to 1 \qquad \frac{(\beta_{rm} + \beta_{wm})P_2}{n} + \frac{\beta_{rm}P_uP_{loc}}{n}$$

$$1, 3 \to 2 \qquad \frac{\beta_{wh}}{n} + \frac{\beta_{wm}P_{loc}}{n}$$

$$4 \to 2 \qquad \frac{(\beta_{rm} + \beta_{wm})P_2}{n} + \frac{\beta_{wh}}{n} + \frac{\beta_{wm}P_{loc}}{n}$$

$$5, 6, 7, 8, 9 \to 2 \qquad \frac{\beta_{wm}P_{loc}}{n}$$

$$s \to 3, s \neq 1 \qquad \frac{\beta_{rm}P_cP_{loc}}{n}$$

$$1 \to 3 \qquad \frac{(K-1)\beta_r}{N} + \frac{\beta_{rm}\eta_cP_{loc}}{n}$$

$$2 \to 4 \qquad \frac{(K-1)\beta_r}{N} + \frac{\beta_{rm}\eta_dP_{loc}}{n}$$

$$s \to 5, s \neq 7 \qquad \frac{\beta_{rm}P_uP_{rem}}{n}$$

$$7 \to 5 \qquad \frac{(\beta_{rm} + \beta_{wm})P_2}{n} + \frac{\beta_{rm}P_uP_{rem}}{n}$$

$$1, 2, 3, 4, 9 \to 6 \qquad \frac{\beta_{wm}P_{rem}}{n}$$

$$5, 7 \to 6 \qquad \frac{\beta_{wh}}{n} + \frac{\beta_{wm}P_{rem}}{n}$$

$$8 \to 6 \qquad \frac{(\beta_{rm} + \beta_{wm})P_2}{n} + \frac{\beta_{wh}}{n} + \frac{\beta_{wm}P_{rem}}{n}$$

$$s \to 7, s \neq 5 \qquad \frac{\beta_{rm}\eta_cP_{rem}}{n}$$

$$5 \to 7 \qquad \frac{(K-1)\beta_r}{N} + \frac{\beta_{rm}\eta_c P_{rem}}{n}$$

$$s \to 8, s \neq 6 \qquad \frac{\beta_{rm}\eta_d P_{rem}}{n}$$

$$6 \to 8 \qquad \frac{(K-1)\beta_r}{N} + \frac{\beta_{rm}\eta_d P_{rem}}{n}$$

$$s \to 9 \qquad \frac{(K-1)w}{N}$$

As an example, the transition $2 \to 1$ covered by the first case occurs when there is a local read miss (β_{rm}) on a memory line which is currently uncached (P_u) and located in the local memory of the requesting processor (P_{loc}). The factor $1/n$ is the probability that the read request maps to the observed cache line. This contains a line in the state 2 before the transition, but this is displaced. Note that remote operations can also induce state transitions locally. For example, the transition $4 \to 2$ can occur if a remote processor performs a miss $(\beta_{rm} + \beta_{wm})$ on a cache line which currently holds a copy of the observed line. The transition occurs when the processor is the only other one holding a copy in the machine. We assume this occurs with probability P_2, the equilibrium probability of there being two sharers in the above Markov model. It can also happen on a write hit (β_{wh}) to the line whereupon the other sharers will be invalidated. The final term $\left(\frac{\beta_{wm}P_{loc}}{n}\right)$ covers the general case transition into state 2 of a write miss on a locally held line. Finally, note that the transition $s \to 9$ corresponds to invalidation— any remote write operation to a line cached locally will cause the line to be invalidated. The factor $(K-1)$ here is the number of remote processors which can issue such a write.

4.4.5. List Operation Probabilities

The probability of each operation occurring in each state is given in Table 4.1. As prescribed, we denote the table by δ so that, for example, $\delta_{2,DC} = (\beta_{rm} + \beta_{wm})P_u$.

4.4.6. Message and Cache/Memory Traffic

To determine the various traffic types it is necessary to understand the operation of the node architecture which is shown in Figure 4.3.

State				Operations			
	CR	AD	RE	DC	DA	DR	IR
1—8	0	0	β_{wh}	$(\beta_{rm} + \beta_{wm})P_u$	$\beta_{rm}P_c$	$\beta_{wm}P_c$	0
9	P_u	$\beta_r P_c$	0	0	0	0	$\beta_w P_c$

Table 4.1: The δ Table

Here there are two buses, one of which can be servicing cache lookups from the processor whilst the other is servicing node memory references from the network. However, the processor may also need to reference memory and/or send messages to one or more remote nodes. Similarly, the network controller may need to access the cache.

There is a simultaneous resource possession problem here although, in fact, only the network controller may hold both buses at the same time to avoid deadlock. If the processor requires access to the memory bus it buffers its request, releases the cache bus and then queues separately for the memory bus. It may buffer either a short transaction (containing no copy of a line, e.g. to update the home memory status or issue a short message) or a long transaction (containing a copy of a cache line, e.g. as a result of write-back following a displacement).

This process, which is represented explicitly in the node model, requires the bus transactions to be partitioned into four Groups (i.e. service classes): those requests which require just the processor bus, those initiated by the processor which require both the cache bus and the memory bus, those which require just the memory bus and, finally, those initiated by the network which require both the memory bus and the cache bus. A Group 2 transaction which requires both buses will be split into two independent transactions, one on each bus. However the memory bus transaction is guaranteed to find the cache bus idle and so will not have to queue for it a second time after acquiring it initially. A Group 4 transaction, on the other hand, may have to contend with the processor for access to the cache bus; its service time at the memory bus is therefore augmented with a queueing time at the cache bus in the model of simultaneous resource possession used.

This particular model uses 24 transactions; 22 different types of bus transaction, a short network message transaction and a long message network transaction. We label the classes C_1, C_2, ..., C_{23} and specify the service time for each in clock cycles. This is given in Table 4.2. Note that each class belongs to exactly one group. We write T_i to be the time (in seconds) for bus transaction C_i to complete once it has been granted the bus(es) required. These times are easily computed by multiplying the bus cycle counts in Table 4.2 by the clock cycle time, t_{clock} say. We also write T_{short} and T_{long} to denote the time taken to send short and long messages respectively.

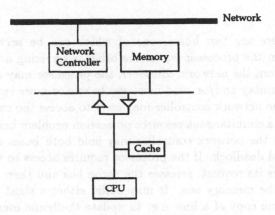

Figure 4.3: ALITE Node Organisation

The number of occurrences of each transaction type are specified in Table 4.3. The first two columns specify the operation and state. The fourth column details the bus transaction classes that are invoked in order to complete the given operation in the given state, together with the number of short and long messages sent. Since each operation/state pair may produce a number of transaction sequences depending on whether a particular line is local/remote, clean/dirty, cached/uncached etc. the corresponding probability is listed separately with each row.

Note that reductions are a special case since the reducing processor sends and receives as many short messages as there are members in the associated sharing list. We estimate the cost of this by using the mean length of a sharing list, Δ, computed from the Markov model of the sharing process.

By substituting the class descriptions from Table 4.2 in place of

Class	Description	Cache bus	Mem bus
Group 1			
C_1	Read from cache	26	0
C_2	Write to cache	8	0
C_3	Buffer short transaction	12	0
C_4	Buffer long transaction	32	0
Group 2			
C_5	Update memory status	3	22
C_6	Ditto and send short	3	30
C_7	Ditto and send long	3	46
C_8	Ditto and read mem to cache	27	89
C_9	Write to mem and read to cache	27	94
C_{10}	Write to mem and send short	3	85
C_{11}	Send short	3	11
C_{12}	Send long	3	27
Group 3			
C_{13}	Update home mem status	0	19
C_{14}	Ditto and send short	0	27
C_{15}	Ditto and send long	0	83
C_{16}	Write to mem and send short	0	75
Group 4			
C_{17}	Invalidate cache	10	13
C_{18}	Read from cache to network	28	31
C_{19}	Write from message to cache	29	32
C_{20}	Transfer mem to cache	26	85
C_{21}	Update mem status & complete	3	22
C_{22}	Complete operation	3	6

Table 4.2: Service classes

the C_i in Table 4.3 a descriptive breakdown of each operation/state pair is produced. For example, the bus and network traffic involved by performing operation DA (Deletion-Addition) in state 2 depends on the status and location of the new line which is to be read. There are four cases, each occurring with an associated probability.

Op.	State	Prob.	Transaction classes invoked
CR	9	P_{loc}	$C_3 + C_8$
		P_{rem}	$C_3 + C_{11} + C_{15} + C_{19} + S + L$
AD	9	P_{loc}	$C_3 + C_8$
		$P_{rem}P_{val}$	$C_3 + C_{11} + C_{15} + C_{19} + S + L$
		$P_{rem}P_{inv}$	$C_3 + C_{11} + C_{14} + C_{15} + C_{19}$
RE	1	1	$C_2 + C_3 + C_5$
	2	1	C_2
	3	1	$C_3 + C_6 + (\Delta - 1)C_{17} + C_{21} + (\Delta)S$
	4	1	$C_3 + C_6 + (\Delta - 1)C_{17} + C_{14} + C_{21}$
	5	1	$C_3 + C_{11} + C_{14} + C_{22} + 2S$
	6	1	C_2
	7,8	1	$C_3 + C_{11} + C_{14} + (\Delta - 1)C_{17} + C_{14} + C_{22} + (\Delta + 3)S$
DC	1,3,4	P_{loc}	$C_3 + C_8$
		P_{rem}	$C_3 + C_6 + C_{15} + C_{19} + S + L$
	2	P_{loc}	$C_4 + C_9$
		P_{rem}	$C_4 + C_{10} + C_{15} + C_{19} + S + L$
	5	P_{loc}	$C_3 + C_{11} + C_{14} + C_{20} + 2S$
		P_{rem}	$C_3 + C_{11} + C_{14} + C_{15} + C_{19} + 2S + L$
	6	P_{loc}	$C_4 + C_{12} + C_{16} + C_{20}$
		P_{rem}	$C_4 + C_{12} + C_{16} + C_{15} + C_{19} + S + 2L$
	7,8	P_{loc}	$C_3 + C_{11} + C_{14} + C_{13} + C_{20} + (\Delta/2 + 5)S$
		P_{rem}	$C_3 + C_{11} + C_{14} + C_{13} + C_{15} + C_{19} + (\Delta/2 + 5)S + L$
DA	1,3,4	$P_{loc}P_{val}$	$C_3 + C_8$
		$P_{loc}P_{inv}$	$C_3 + C_6 + C_{18} + C_{19} + S + L$
		$P_{rem}P_{val}$	$C_3 + C_6 + C_{15} + C_{19} + S + L$
		$P_{rem}P_{inv}$	$C_3 + C_6 + C_{14} + C_{18} + C_{19} + 2S + L$
	2	$P_{loc}P_{val}$	$C_4 + C_9$
		$P_{loc}P_{inv}$	$C_4 + C_{10} + C_{18} + C_{19} + S + L$
		$P_{rem}P_{val}$	$C_4 + C_{10} + C_{15} + C_{19} + S + L$
		$P_{rem}P_{inv}$	$C_4 + C_{10} + C_{14} + C_{18} + C_{19} + 2S + L$
	5	$P_{loc}P_{val}$	$C_3 + C_{11} + C_{14} + C_{20} + 2S$
		$P_{loc}P_{inv}$	$C_3 + C_{11} + C_{14} + C_{14} + C_{18} + 3S + L$
		$P_{rem}P_{val}$	$C_3 + C_{11} + C_{14} + C_{15} + C_{19} + 3S + L$
		$P_{rem}P_{inv}$	$C_3 + C_{11} + C_{14} + C_{14} + C_{18} + C_{19}$
	6	$P_{loc}P_{val}$	$C_4 + C_{12} + C_{16} + C_{20} + S + L$
		$P_{loc}P_{inv}$	$C_4 + C_{12} + C_{14} + C_{14} + C_{18} + 3S + L$
		$P_{rem}P_{val}$	$C_3 + C_{11} + C_{14} + C_{15} + C_{19} + 3S + L$
		$P_{rem}P_{inv}$	$C_3 + C_{11} + C_{14} + C_{14} + C_{18} + C_{19}$
	7,8	$P_{loc}P_{val}$	$C_3 + C_{11} + C_{14} + C_{13} + C_{20} + (\Delta/2 + 5)S$
		$P_{loc}P_{inv}$	$C_3 + C_{11} + C_{14} + C_{13} + C_{14} + C_{18} + C_{19} + (\Delta/2 + 6)S + L$
		$P_{rem}P_{val}$	$C_3 + C_{11} + C_{14} + C_{13} + C_{15} + C_{19} + (\Delta/2 + 6)S + L$
		$P_{rem}P_{inv}$	$C_3 + C_{11} + C_{14} + C_{13} + C_{14} + C_{18} + C_{19} + (\Delta/2 + 7)S + L$

Table 4.3: ALITE bus and memory transactions

Op.	State	Prob.	Transaction classes invoked
DR	1,3,4	$P_{loc}P_u$	$C_3 + C_8$
		$P_{loc}P_c$	$C_3 + C_6 + \Delta C_{17} + C_{20} + (\Delta + 1)S$
		$P_{rem}P_u$	$C_3 + C_6 + C_{15} + C_{19} + S + L$
		$P_{rem}P_c$	$C_3 + C_6 + C_{14} + \Delta C_{17} + C_{18} + C_{19} + (\Delta + 2)S + L$
	2	$P_{loc}P_u$	$C_4 + C_9$
		$P_{loc}P_c$	$C_4 + C_{10} + \Delta C_{17} + C_{18} + C_{19} + (\Delta + 1)S$
		$P_{rem}P_u$	$C_4 + C_{10} + C_{15} + C_{19} + S + L$
		$P_{rem}P_c$	$C_4 + C_{10} + C_{14} + \Delta C_{17} + C_{18} + C_{19} + (\Delta + 2)S + L$
	5	$P_{loc}P_u$	$C_3 + C_{11} + C_{14} + C_{20} + 2S$
		$P_{loc}P_c$	$C_3 + C_{11} + C_{14} + C_{14} + \Delta C_{17} + C_{20} + (\Delta + 3)S$
		$P_{rem}P_u$	$C_3 + C_{11} + C_{14} + C_{15} + C_{20} + 3S + L$
		$P_{rem}P_c$	$C_3 + C_{11} + C_{14} + C_{14} + \Delta C_{17} + C_{15} + C_{19} + (\Delta + 4)S + L$
	6	$P_{loc}P_u$	$C_4 + C_{12} + C_{16} + C_{20} + S + L$
		$P_{loc}P_c$	$C_4 + C_{12} + C_{16} + C_{14} + \Delta C_{17} + C_{20} + (\Delta + 2)S + L$
		$P_{rem}P_u$	$C_4 + C_{12} + C_{16} + C_{15} + C_{19} + 2S + 2L$
		$P_{rem}P_c$	$C_4 + C_{12} + C_{16} + C_{14} + \Delta C_{17} + C_{15} + C_{19} + (\Delta + 3)S + 2L$
	7,8	$P_{loc}P_u$	$C_3 + C_{11} + C_{14} + C_{13} + \Delta C_{17} + C_{20} + (\Delta/2 + 5)S$
		$P_{loc}P_c$	$C_3 + C_{11} + C_{14} + C_{13} + C_{14} + C_{18} + C_{19} + (3\Delta/2 + 6)S$
		$P_{rem}P_u$	$C_3 + C_{11} + C_{14} + C_{13} + C_{15} + C_{19} + (\Delta/2 + 6)S + L$
		$P_{rem}P_u$	$C_3 + C_{11} + C_{14} + C_{13} + C_{14} + C_{18} + C_{19} + (3\Delta/2 + 7)S + L$
IR	9	$P_{loc}P_u$	$C_3 + C_8$
		$P_{loc}P_c$	$C_3 + C_{11} + \Delta C_{17} + C_{20} + (\Delta + 1)S$
		$P_{rem}P_u$	$C_3 + C_{11} + C_{15} + C_{19} + S + L$
		$P_{rem}P_c$	$C_3 + C_{11} + C_{14} + \Delta C_{17} + C_{18} + C_{19} + (\Delta + 2)S + L$
RH	1—8	1	C_1

Table 4.3: ALITE bus and message transactions (cont.)

For example, if the line being read is on the same node as the line being displaced and is valid with respect to other sharers ($P_{loc}P_{val}$) then the operation can be completed without communication: the dirty line being displaced must be written back to memory (involving claiming both the cache and memory buses) after which the new line can be read into the cache in the opposite direction. This requires one transaction on each bus ($C_4 + C_9$) for which there is an associated total delay (see Table 4.2) of 59 cycles on the cache bus and 94 cycles on the memory bus, with 27 cycles of the latter being consumed with both buses held.

If, however, the new line's home node is elsewhere and if the home copy is invalid, due to a write operation, then more work must be done: the dirty line being displaced must be written back to memory. A long transaction is issued on the cache bus and a subsequent transaction on the memory bus writes the line to memory and sends a short message to the home node. When this is received the home node updates the pointer status field of the addressed line (adding the new node to the sharing list) and forwards a short request message to the current sharing list head which has an up-to-date copy of the line. This node claims its memory *and* cache buses, fetches a copy of the line from its cache and returns it as a long message, targeted to the initiator of the DA operation. When this long message is received both buses at the initiator are claimed and the line is transferred to the cache; this in turn restarts the processor. In this latter case two short messages and one long message are sent–their transmission times do not affect the bus queueing times but do add delays to the overall read/write response time.

The (sum of the) coefficient(s) of C_i for operation p in state s, multiplied by the associated probabilities, determines the expected number of occurrences of bus transaction $i, 1 \le i \le 22$, on a memory access. Similarly, the numbers of short and long messages.

4.4.7. Modelling the Node

We give here a brief overview of the queueing model of the node, which has been documented in full elsewhere[12].

The node architecture is modelled as a queueing network with a server representing each bus. The bus delays depend on the type of transaction and so the transaction classes in Table 4.2 become service classes in an M/G/1 queueing model. Pointer traversal and pointer maintenance are handled by the network controller; this is pipelined and the associated delays are therefore assumed to be subsumed by the message transmission times.

The queueing network is complicated by the fact that internal requests from the processor and external requests from the network may require either one bus, or both buses, to complete a transaction. This leads to a form of simultaneous resource possession in the queueing network and hence blocking-before-service. We utilise an approximate solution to this problem by augmenting the service time at the memory bus with the waiting time at the cache bus, for the Group 4 bus transaction classes; Group 2 is similar except that there is no queueing for the cache bus, as previously described.

The response time of the memory system, R, may now be now obtained from the various bus transaction times and the short and long message transmission times t_{short} and t_{long}. Defining $\Omega = \{CR, ..., RH\}$ to be the set of coherency operations and $S = \{CXC, ..., INV\}$ to be the set of sharing list operations, we obtain:

$$R = \sum_{p \in \Omega} \sum_{s \in S} q_s \, \delta_{s,p} \left(\left(\sum_{i=1}^{C} T_{s,p,i} \, w_i \right) + (S_{s,p} \, t_{short} + L_{s,p} \, t_{long}) \right)$$

where w_i is the total waiting time for class i bus transactions. These are computed from an M/G/1 model of the dual bus system, suitably modified to accommodate the simultaneous resource possession. The full analysis uses the generating function of the steady-state queue length distribution to determine the respective queuing times. Processor utilisation is then:

$$\pi = \frac{1}{1 + \tau R}$$

Note, however, that in calculating the arrival rates for each bus transaction class we assumed the existence of π by virtue of the factor $\tau = \pi \tau_0$ in the defining summation. We have thus introduced a new fixed-point problem for determining π and again appeal to an iterative solution method. We note in passing that some care has to be taken in updating the approximation to π in order to ensure convergence.

4.5. CASE STUDY—SCI

The SCI protocol[4,5] is an IEEE standard protocol based on *bidirectionally* linked sharing lists. Although there are some similarities with the ALITE protocol above, it differs significantly in the line states and in the coherency management messages required.

In contrast to the ALITE model, the model here employs a very simple model of the node but a more sophisticated model of the communication network, which is a ring (the default for SCI). It also illustrates a more sophisticated workload model which distinguishes three different classes of data. This is instructive since it demonstrates how details of this kind can be incorporated by suitable enhancement to the line states and associated Markov process.

4.5.1. Line States

A memory line may be in two possible states

1 Home The line is not cached by any processor

2 Cached The line is cached by one or more processors

The caches hold local copies of memory lines and these copies may either be *clean* or *dirty*. The workload model here distinguishes *private* data from *shareable* data. Private data is typically stored local to the processor although this locality is formally specified by a parameter σ which denotes the probability that a private line resides in the local memory. This enables non-local private data to be modelled; such data will be cached in state 1 or 2. There are eleven basic line states:

1 Private Clean The location contains a clean copy of private (non-shareable) data

2 Private Dirty The location contains a dirty copy of private (non-shareable) data

3 Only Clean The location contains a clean copy of a memory line and is alone in the sharing-list.

4 Only Dirty The location contains a dirty copy of a memory line and is again alone in the sharing list.

5 Head Clean As 3 but the location is at the head of a list containing at least two members.

6 Head Dirty As 5 but the cached copy is dirty. All members of the list hold the same dirty copy, but this is different to the home memory line.

7 Mid Clean The location is in the middle of the list (i.e. at neither the head nor tail in a list with at least three members); the cached copy is clean.

8 Mid Dirty The location is in the middle of the list; the cached copy is dirty.

9 Tail Clean As 5 but the location is at the tail of the sharing list.

10 Tail Dirty As 6 but the location is at the tail of the sharing list.

11 Invalid The location contains no usable information.

The workload model assumes that there are m special cache lines which may contain shareable control variables, e.g. locks, in addition to ordinary shared data and non-local private data, whilst others may contain only the last two. The cache regions are numbered I and II respectively and are considered separately in the model. We make the assumption that the memory contains exactly m control lines in total such that there can be no displacement of one control line by another. Thus we assume that different types of data do not co-exist on the same cache line; this is easily ensured by a suitable allocation of global memory addresses. Note that the entries of a sharing list must all be in the same region of their corresponding caches so that within each region the cache lines are statistically identical. There is thus a separate model for each region. The parameters γ_1, γ_2 and γ_3 denote the probability that a memory reference is to a control line, an ordinary shared data line, and a prviate line respectively.

The cache line states in Region I are therefore augmented to distinguish two types of shareable data; states 3–10 are annotated with the subscript 'a' if the (shared) line contains control information and 'b' if it contains ordinary shared data. No annotations are required for Region II. Thus, for example, state 5_a indicates a clean line forming the head of a sharing list which contains a control variable. The cache lines in Region I therefore have nineteen possible states, and those of Region II eleven.

4.5.2. Coherency Operations

The coherency operations are identical to those in the ALITE architecture although there are significant differences in the way some of them are implemented. This is due in part to the fact that the sharing list here is doubly linked. There are other differences, however. For example, in order to write a line the write must, from the definition of the protocol, be at the head of the sharing list before the list can be reduced. This introduces additional traffic in some cases. As before, we label the operations {CR,...,IR}.

4.5.3. Sharers

We summarise the model for Region II since it involves fewer states and transitions than Region I. The Markov process for the number of sharers of lines that map to Region II is shown in Figure 4.4. The structure of the model is similar to that of ALITE but lists of length i, $1 \leq i \leq K - 1$, are distinguished to indicate whether or not they include the cache line at the home node. State K represents a list of length K which must include the home node and state 0 represents an uncached line. There are transitions between adjacent list states that include or exclude the home node as well as transitions between these two subsets of states caused by the home node joining or jumping off a sharing list. There are similar Markov models for lines that map to Regions I_a and I_b.

These models are solved to obtain the equilibrium probability distribution of the sharing list length, together with its mean value, Δ_a, Δ_b, Δ.

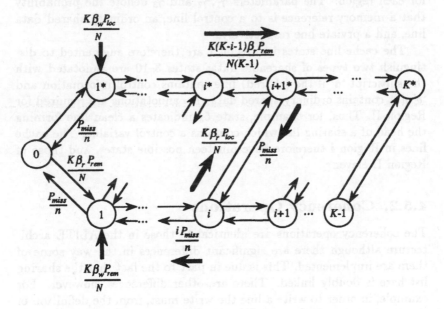

Figure 4.4: Markov model of the number of SCI line sharers

4.5.4. Transition Rates

We show below the transition rates for the cache lines in Region II. The rates for Region I can be produced similarly although there are more of them due to the additional states within this region. The rates are defined in terms of η' which is the probability that a given line is cached somewhere, and ϵ' which is the probability that a cached line is in a clean state. These are estimated from the probability that a sharing list is empty and the line state probabilites respectively (at equilibrium), which again leads to a fixed point problem for the $q'_j, 1 \leq j \leq 11$ in Region II and $q_j, 1 \leq j \leq 19$ in Region I similarly. The rates for Region II are:

$$2 \to 1, s \neq 11 \qquad \frac{\alpha(1 - \beta_m)\gamma_3}{n}$$

$$s \to 1, s \neq 1, 2 \qquad \frac{\alpha\gamma_3}{n}$$

$$s \to 2 \qquad \frac{(1 - \alpha)\gamma_3}{n}$$

$$s \to 3, s \neq 1, 2, 5, 9, 11 \qquad \frac{\alpha(1 - \beta_m)(1 - \eta)\gamma_2}{n}$$

$$1, 2, 11 \to 3 \qquad \frac{\alpha(1 - \eta)\gamma_2}{n}$$

$$5 \to 3, 9 \to 3 \qquad \frac{\alpha(1 - \beta_m)(1 - \eta)\gamma_2}{n} + \frac{p'_3(1 - \beta_m)}{n}$$

$$s \to 4, s \neq 6, 10 \qquad \frac{(1 - \alpha)\gamma_2}{n}$$

$$6 \to 4, 10 \to 4 \qquad \frac{(1 - \alpha)\gamma_2}{n} + \frac{p'_3(1 - \beta_m)}{n}$$

$$s \to 5, s \neq 1, 2, 7, 11 \qquad \frac{\alpha(1 - \beta_m)\eta\epsilon\gamma_2}{n}$$

$$1, 2, 11 \to 5 \qquad \frac{\alpha\eta\epsilon\gamma_2}{n}$$

$$7 \to 5 \qquad \frac{\alpha(1 - \beta_m)\eta\epsilon\gamma_2}{n} + \frac{p'_1(1 - \beta_m)}{n}$$

$$s \to 6, s \neq 1, 2, 8, 11 \qquad \frac{\alpha(1 - \beta_m)\eta(1 - \epsilon)\gamma_2}{n}$$

$$1, 2, 11 \to 6 \qquad \frac{\alpha\eta(1 - \epsilon)\gamma_2}{n}$$

$$8 \to 6 \qquad \frac{\alpha(1 - \beta_m)\eta(1 - \epsilon)\gamma_2}{n} + \frac{p'_1(1 - \beta_m)}{n}$$

$$5 \to 7, 6 \to 8 \qquad \frac{(K - \Delta')\alpha\gamma_2}{N - m}$$

$$3 \to 9, 4 \to 10 \qquad \frac{(K-1)\alpha\gamma_2}{N-m}$$

$$7 \to 9, 8 \to 10 \qquad \frac{p_2'(1-\beta_m)\gamma_2}{n}$$

$$s \to 11, s \neq 1, 2 \qquad \frac{(K-1)(1-\alpha)\gamma_2}{N-m}$$

Δ' is the mean length of a sharing list in Region II.

4.5.5. List Operation Probabiities

The probability of each coherency operation occurring in each state in Region II is given in Table 4.4. We write this δ', remarking that a similar table δ exists for Region I. Note that the notation $\overline{x}$ denotes $1 - x$.

State	CR	AD	DC	DA	DR	TR	HR	IR
				Operation				
1,2	0	0	$\overline{\eta'}\,\overline{\beta_m}$	$\alpha\eta'\,\overline{\beta_m}$	$\overline{\alpha}\eta'$	0	0	0
3,4	0	0	$\overline{\eta'}\,\overline{\beta_m}$	$\alpha\eta'\,\overline{\beta_m}$	$\overline{\alpha}\eta'$	0	0	0
5,6	0	0	$\overline{\eta'}\,\overline{\beta_m}$	$\alpha\eta'\,\overline{\beta_m}$	$\overline{\alpha}\eta'$	0	$\overline{\alpha}\beta_m$	0
7,8	0	0	$\overline{\eta'}\,\overline{\beta_m}$	$\alpha\eta'\,\overline{\beta_m}$	$\overline{\alpha}\eta'$	$\overline{\alpha}\beta_m$	0	0
9,10	0	0	$\overline{\eta'}\,\overline{\beta_m}$	$\alpha\eta'\,\overline{\beta_m}$	$\overline{\alpha}\eta'$	$\overline{\alpha}\beta_m$	0	0
11	$1-\eta'$	$\alpha\eta'$	0	0	0	0	0	$\overline{\alpha}\eta'$

Table 4.4: The δ' Table

4.5.6. Message and Cache/Memory Traffic

The model here uses a straightforward queueing model of each node, but incorporates a more sophisticated model of the network than the previous case study. Node traffic is divided into two classes for simplicity, namely traffic directed toward a node cache and traffic directed toward memory. Essentially the bus transaction classes which were detailed individually in the ALITE study have been aggregated in this model.

The network has a unidirectional ring architecture. We again identify two classes of message (short and long messages as above) but here use their predicted frequencies to generate inputs to an M/G/1 model of the ring. Table 4.5 shows the total cache, memory and message traffic generated at a node by each operation in each state. Since there are only two transaction classes for the node we list them explicitly in the table (c.f. Table 4.3 referred to above) which defines the non-zero entries of six new tables S, S', L, C, C' and M representing the number of short messages (Region I and Region II respectively), long messages, cache accesses (Region I and Region II respectively) and memory accesses.

Op.	State	Short		Long	Cache		Mem
		I $(r=a,b)$	II		I $(r=a,b)$	II	
		(S)	(S')	(L)	(C)	(C')	(M)
CR	11	1^*	1^*	1^*	2	2	1
AD	11	1^*+1	1^*+1	1	3	3	1
DC	1	1^*	1^*	1^*	2	2	1
	2	1^*	1^*	1^*+1	2	2	2
	3	2^*	2^*	1^*	2	2	2
	4	1^*	1^*	2^*	2	2	2
	5,6	2^*+1	2^*+1	1^*	3	3	2
	7,8	1^*+2	1^*+2	1^*	4	4	1
	9,10	1^*+1	1^*+1	1^*	3	3	1
DA	1	1^*+1	1^*+1	1	3	3	1
	2	1^*+1	1^*+1	2	3	3	2
	3	2^*+1	2^*+1	1	3	3	2
	4	1^*+1	1^*+1	$1^*\,|\,1$	0	3	2
	5,6	2^*+2	2^*+2	1	4	4	2
	7,8	1^*+3	1^*+3	1	5	5	1
	9,10	1^*+2	1^*+2	1	4	4	1
DR	3	$2^*+1+2\Delta_M$	$2^*+1+2\Delta'_M$	1	$3+\Delta_M$	$3+\Delta'_M$	2
	4	$1^*+1+2\Delta_M$	$1^*+1+2\Delta'_M$	1^*+1	$3+\Delta_M$	$3+\Delta'_M$	2
	5,6	$2^*+2+2\Delta_M$	$2^*+2+2\Delta'_M$	1	$4+\Delta_M$	$4+\Delta'_M$	2
	7,8	$1^*+3+2\Delta_M$	$1^*+3+2\Delta'_M$	1	$5+\Delta_M$	$5+\Delta'_M$	1
	9,10	$1^*+2+2\Delta_M$	$1^*+2+2\Delta'_M$	1	$4+\Delta_M$	$4+\Delta'_M$	1
TR	7,8	$1^*+3+2(\Delta^r_H - 1)$	$1^*+3+2(\Delta'_H - 1)$	0	$3+\Delta^r_H$	$3+\Delta'_H$	1
	9,10	$1^*+2+2(\Delta^r_H - 1)$	$1^*+2+2(\Delta'_H - 1)$	0	$2+\Delta^r_H$	$2+\Delta'_H$	1
HR	5,6	$2(\Delta^r_H - 1)$	$2(\Delta'_H - 1)$	0	Δ^r_H	Δ'_H	0
IR	11	$1^*+1+2\Delta_M$	$1^*+1+2\Delta'_M$	1	$3+\Delta_M$	$3+\Delta'_M$	1

Table 4.5: SCI cache, memory and message transactions

Note that some messages may by-pass the ring, specifically if they are directed toward the same node as the initiator. These are labelled with a '*'. Shareable data is assumed to be uniformly distributed across the nodes of the machine. In the table, therefore, the term x^* can be read as $x(K-1)/K$.

Note that reductions are a special case since the reducing processor sends and receives as many short messages as there are members in the sharing list behind its cached block. We estimate this by using the mean length of a non-singleton sharing list within each region. We require five values here depending on whether or not the writer is already a member of the sharing list to be reduced. Δ'_H denotes the mean length of a sharing list of which the writer is already a part. Δ'_M denotes the mean length of a sharing list prior to the writer adding itself as a result of a write miss. Thus, $\Delta'_M \leq K - 1$ whereas $\Delta'_H \leq K$. Similar quantities are defined for Region I appropriately annotated with the line type (a or b). Note that when joining a sharing list in Region I for the purposes of reduction we cannot determine which line type we are joining. We therefore define Δ_M to be a weighted average of Δ^a_M and Δ^b_M in this case. The various Δ values are derived from the Markov models of the sharing processes in each Region/area.

4.5.7. Modelling the Node

The ring model assumes that a short message issued by a transmitting node will perform one full circuit of the ring (i.e. through $K - 1$ ring buffers). For SCI in practice, the destination forwards an *echo* packet on receipt of a message. However, these are of similar length to short messages, so we model it as though the short message makes a full circuit of the ring. A long message is converted to a short echo message once it has reached the receiving node.

Messages originating from the ring have priority over messages in the transmit queue originating from the node so the model uses Cobham's formulae to determine the mean waiting times of messages in the transmit queue and ring buffer. Table 4.5 is first used to determine the rate at which messages are generated by the transmit queue and ring buffer of each node:

$$\lambda_s = \tau(P_I \sum_{s,a} q_s \, \delta_{s,a} \, S_{s,a} + P_{II} \sum_{s',a} q'_{s'} \, \delta'_{s',a} \, S'_{s',a})$$

$$\lambda_l = \tau(P_I \sum_{s,a} q_s \, \delta_{s,a} \, L_{s,a} + P_{II} \sum_{s',a} q'_{s'} \, \delta'_{s',a} \, L_{s',a})$$

where $\tau = \pi \tau_0$ is the rate at which memory requests are submitted by a processor, and where P_I and P_{II} represent the probability that a memory reference addresses the cache in Regions I and II respectively:

$$P_I = \gamma_1 + (\gamma_2 + \gamma_3(1 - \sigma))\frac{m}{n}$$

$$P_{II} = (\gamma_2 + \gamma_3(1 - \sigma))\frac{n - m}{n}$$

Since the length of a long message is a multiple M of the length of a short message the various transmission times can be expressed in terms of the transmission time of a short message, denoted by t_{short}.

The M/G/1 model uses the first and second moments of the service time in order to compute the mean queueing time at the ring buffer and transmit queue, Q_r and Q_t respectively. This straightforward analysis finally yields the mean transmission time of short and long messages, respectively:

$$T_s = Kt_{short} + Q_t + (K - 1)Q_r$$

$$T_l = (K/2)(1 + M)t_{short} + Q_t + (K - 1)Q_r$$

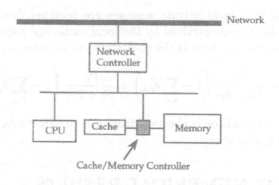

Figure 4.5: SCI Node Architecture

The cache/memory controller in Figure 4.5 is represented by a single shared queue. The requests are generated both by the processor(s) and the SCI ring via the network controller and contain a mixture of cache and memory requests. We distinguish the two in the model which allows changes in the ratio of cache speed to memory speed to be explored.

We can determine the mean number of cache and memory accesses (n_c and n_m respectively) for each action in a similar manner to the number of short and long messages, as given above. The number of cache and memory accesses per operation/state are given in the tables C, C' and M defined earlier and from these we can calculate the rate at which cache and memory accesses are produced by a processor (λ_c and λ_m repectively):

$$\lambda_c = \tau(1 - \gamma_3\sigma) + \lambda_t\frac{n_c}{n_c + n_m}$$

$$\lambda_m = \tau\gamma_3\sigma + \frac{\tau(\gamma_1 + \gamma_2)(1 - \beta_m)}{K} + \lambda_t\frac{n_m}{n_c + n_m}$$

The total arrival rate of cache and memory accesses is $\lambda_{cm} = \lambda_c + \lambda_m$ which is used to compute the mean queuing time at the cache/memory controller, Q_{cm} using the Pollaczek-Khinchine formula.

We finally obtain the mean time to service a memory request:

$$R = (1 - \sigma\gamma_3)[\frac{\lambda_s T_s + \lambda_l T_l}{\tau(1 - \sigma\gamma_3)} + n_c(Q_{cm} + t_{cache}) + n_m(Q_{cm} + t_{mem})$$
$$+ \ p_{hit}\,(Q_{cm} + t_{cache})] + \sigma\gamma_3(Q_{cm} + t_{mem})$$

remembering that local private accesses are serviced directly by the memory, cache hits are serviced by the local cache and cache misses by the SCI protocol. p_{hit} here is the probability of a cache hit:

$$p_{hit} = \frac{P_I}{P_I + P_{II}}\left(1 - \sum_{s,a}\delta_{s,a}\right) + \frac{P_{II}}{P_I + P_{II}}\left(1 - \sum_{s',a}\delta'_{s',a}\right)$$

The processor utilisation is obtained from the same formula as was used in the ALITE case study.

4.6. SOME NUMERICAL RESULTS

The two models described were implemented in Mathematica 2.2 and executed on a Power Macintosh 7100/66. Each computed the equilibrium probability distributions for the line state probabilities and the number of sharers of memory lines, and hence node utilisation, system throughput and average memory latency. Numerous experiments are possible and many have been carried out to predict, for example, the quantitative effects on performance of varying cache hit rates and locality of reference. The latter includes the proportion of memory accesses to the system variables that control data structures in the SCI model, represented by the parameter γ_1. Here we consider the processor utilisation and average memory latency for the ALITE and SCI models with parameters chosen based on their specifications together with typical, observed workload characteristics. As a baseline parameterisation, we assume uniform memory access, i.e. a memory access addresses each node with equal probability. Although this is not realistic, it provides a benchmark against which to compare alternative architectures and also generates more sharing with which to test the model than a locality-tuned, real application.

4.6.1. ALITE

Graphs of node utilisation and memory latency against the number of nodes (up to 32) were plotted for various memory sizes; 1, 10, 100 and 1000 times the node cache size (Figures 4.6 and 4.7). These suggest that the architecture scales well, especially with large memory, in that utilisation stays high and even memory latency only increases by a few percent, except with relatively small memory size. The implication is that larger memory results in fewer invalidations and hence cleaner lines and less sharing list maintenance. This is a somewhat optimistic interpretation in that the hit ratio was *fixed* at 0.9. Hence the decrease in the number of invalidations was not counterbalanced by an increased miss rate which might be expected from scaling up an application. Nevertheless, a not inconsiderable benefit in exploiting locality (giving a high hit ratio) is indicated.

4.6.2. SCI

The performance of the SCI protocol (Figures 4.8 and 4.9) degrades considerably as the number of nodes increases, utilisation falling below 0.2 in a ring of 32 processors at small memory sizes. Whilst not an entirely realistic scenario, this result indicates the serious overhead that can result when sharing lists grow. It is the contention in ring buffers that prevents a ring interconnect from scaling linearly, and here we see the effect of significantly greater than zero mean queue lengths— around 4 for a 16-node ring and over 5 for 32 nodes. Notice that sharing list lengths do not increase dramatically (actually close to logarithmically) with the number of nodes. This is because of a significant write probability (0.2) which periodically resets the length of a list to one.

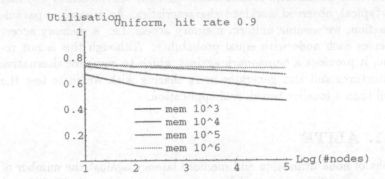

Figure 4.6: Processor utilisation curves for the ALITE protocol

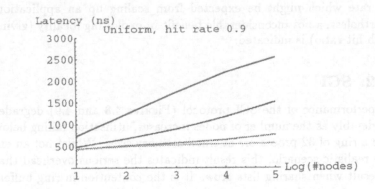

Figure 4.7: Memory latency curves for the ALITE protocol

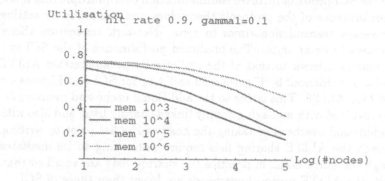

Figure 4.8: Utilisation curves for the SCI protocol

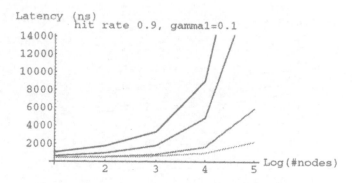

Figure 4.9: Memory latency for the SCI protocol

4.6.3. Comparison of the protocols

Since the SCI protocol involves communication over multiple ring links, the performance of the protocols themselves was compared by setting the message transmission times to zero. Network contention therefore ceases to be an issue. The predicted performance of the SCI system remains inferior to that of the realistically parameterised ALITE model as is evidenced by Figures 4.10 and 4.11 for SCI and Figures 4.6 and 4.7 for ALITE. This is due to the additional cache and memory accesses involved with managing doubly linked sharing lists, and also with the additional overhead of taking the head of the list prior to writing. Although the ALITE sharing lists require unhooking to be mediated through the home node, in practice the sharing lists are small so that, in fact, the ALITE protocol overheads are lower than those of SCI.

When message transmission times are set to zero, the ALITE model scales better, although the curves for the smallest memory size still degrade significantly as the number of nodes increases. In fact, they are not so different from those of Figures 4.6 and 4.7. This is due mainly to the dominance of the bus-contention arising from increasing invalidation traffic.

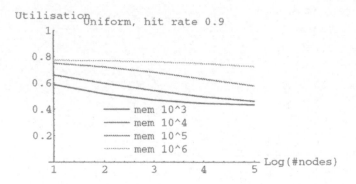

Figure 4.10: Utilisation curves for SCI with $t_{short} = t_{long} = 0$

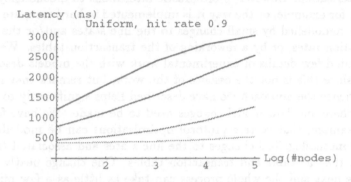

Figure 4.11: Memory latency curves for the SCI protocol with
$t_{short} = t_{long} = 0$

4.7. SUMMARY AND CONCLUSIONS

We have described a general methodology for modelling the performance of distributed shared-memory multiprocessors, focusing in particular on the modelling of the cache memory system. The methodology breaks down into six steps, the first five of which are specific to the coherency protocol and the last of which relates to the modelling of the procesing nodes and communication network.

We have found the technique to be applicable to a range of distributed shared memory systems and have illustrated its use here with two case studies. The two examples use different coherency protocols and have different node and network architectures. Between them they demonstrate the flexibility of the method, in particular in coping with radically different node and network models. An arbitrary level of detail relating to the memory, cache, node bus and network traffic can be captured in the form of transaction class tables which collectively detail the amount of "work" required to perform each coherency operation in each state.

One of the benefits of a stochastic model in this context is the relative ease with which the effect on performance of various design changes, or alterations to the assumed workload, can be analysed. For straightfoward changes, this is often simply a case of changing the model parameterisation. However, more drastic alterations to a coherency protocol, for example, or the way it is implemented in hardware, are often easily remodelled by small changes to the line states and/or the state transition rates, or by a reworking of the transaction tables. We have presented few details of experimental work with the models described here since this is not the essence of the work, but remark that in our experience the approach we have described helps significantly to identify where and how model changes need to be made. We have found, for example, that certain protocol optimisations can be modelled by small methodogolical changes to the line states and associated transitions, the "δ" table and transaction tables. One change neatly leads to the next and the whole process can take as little as a few minutes from start to finish. The benefits to a designer of the type of system in question are self evident.

References

[1] R. Hexsel and N. Topham, "Performance of SCI Memory Hierarchies", In Proceedings of 8th International Workshop on Support for Large-Scale Memory Architectures, April 1994.

This paper summarises some of the work described in the first author's PhD thesis, including the development of an execution-driven simulation model of an SCI-based multiprocessor. The simulator code does not, however, model the SCI ring explicitly.

[2] Wolfram Research. *Mathematica - A System for Doing Mathematics by Computer*, Wolfram research, 1994.

[3] A.G. Greenberg and I.Mitrani "Analysis of Snooping Caches". *Proc. of Performance 87, 12th Int. Symp. on Computer Performance*, Brussels, December 1987.

This is a classic paper on coherency protocol modelling and one which inspired much of the work in this paper. It develops models for a range of bus-based protocols based on Markov models for the line states. A comparitive analysis of the various protocols is undertaken and the results are validated against a simulation model.

[4] S. Gjessing, D.B. Gustavson, J.R. Goodman, D.V. James, E.H Kristiansen. "The SCI Cache Coherence Protocol". In *Scalable Shared Memory Multiprocessors*, M. Dubois and S. Thakkar, eds., Kluwer academic Publishers, Norwell, Mass. 1992

This paper presents an overview of the SCI protocol. It is essentially a condensed summary of the full IEEE specification document cited below, but greatly abstracted to explain the key principles whilst avoiding low-level implementation details.

[5] The IEEE. "IEEE P1596 Standard Specification". IEEE Publication, 1989.

This is the full IEEE specification for SCI. The specification is accompanied by a C program code which collectively constitutes a chip-level simulation of an SCI node. The specification code is very low level, however, and this makes it unsuitable as the basis for a practical SCI simulation model.

[6] S.L. Scott, J.R. Goodman and M.K. Vernon. "Performance of the SCI ring". *In Proc. of the 19th Annual Int. Sym. on Computer Architecture*. May 1992.

This paper is one of the first papers to address analytical modelling study of SCI. It presents a detailed model of the SCI ring, which is represented by an M/G/1 queue. The paper does not model coherency traffic, however.

[7] A. Saulsbury, T. Wilkinson, J. Carter, and A. Landin, "An Argument For Simple COMA", *In Proc. First IEEE Symposium on High Performance Computer Architecture*, Rayleigh, North Carolina, USA, pp. 276-285, January 1995.

This paper is principally concerned with a new proposal for a coherent cache-only memory architecture (COMA), but includes in its discussions a reference model of a CC-NUMA protocol, ALITE, which has been referred to in this paper. The main idea behind the COMA architecture proposed is to use existing TLB harwdare for virtual memory management to perform page-level associative memory look-up.

[8] A. Nowatzyk *et al*, "The S3.mp Scalable Shared Memory Multiprocessor", *in Proc. of the 1995 International Conference on Parallel Processing*", Oconomowoc, Wisconsin, August 1995, pp. I1–I10.

This paper won the best paper award at the 1995 International Conference on Parallel Processing. It summarises the architecture of Sun's proposals for workstation networks supporting distributed shared memory using additional controller cards and an interface to an optical communication network. The coherency protocol supported has some similarities with the ALITE protocol referred to in this paper.

[9] P.G. Harrison and N.M. Patel. *Performance Modelling of Communication Networks and Computer Architectures*. Addison-Wesley. 1993.

This is a detailed and comprehensive text book covering a wide range of state-of-the-art techniques in analytical performance modelling.

[10] E.D. Lazowska, J. Zahorjan, G.S. Graham and K. Sevcik, "Quantitative System Performance". Prentice Hall. 1984.

This is a more practically oriented, yet very comprehensive, exposition of the state-of-the-art at the date of publication.

[11] M. Thapar and B. Delagi, "Stanford Distributed-Directory Protocol", IEEE Computer, Vol.23, No. 6, June 1990, pp. 78-80.

This paper describes Stanford's own coherency protocol, which is similar in many ways to the ALITE protocol described here. It serves as an excellent reference model as well as being a functional protocol in its own right.

[12] A.J. Bennett, A.J. Field and P.G. Harrison, "Modelling and Validation of Shared Memory Coherency Protocols", Internal Report, Department of Computing, Imperial College, 1996.

This provides a detailed description of the model of the ALITE coherency protocol referred to. Also included is a preliminary model validation against an execution-driven simulation model of the ALITE memory system.

CHAPTER 5

TRACING NONDETERMINISTIC PROGRAMS ON SHARED MEMORY MULTIPROCESSORS

Zulah K. F. Eckert and Gary J. Nutt

5.1. INTRODUCTION

An *event trace* is a sequence of event occurrences transpiring during the execution of a program on a particular execution architecture. That is, when a program is executed by a particular runtime system with a particular operating system on a particular hardware platform, the effect of the execution can be saved as a sequence of event occurrences observed during the execution. Such a trace is collected by selecting a program, data, and execution architecture; instrumenting the system; then collecting and storing the event sequence for subsequent analysis (see Figure 5.1).

Event tracing has been a fundamental technique for computer performance evaluation since the 1960s. Besides their use for post mortem analysis, event traces gained widespread use as a means to drive simulations using a realistic job load. The rationale was that stochastic representations of the load due to a program execution could not capture the full subtlty of a program behavior, since these "synthetic programs" might introduce behavioral localities that did not exist, or not accurately represent those that do exist. The idea of executing a program with a data set on one computer, tracing its behavior, than using the trace to drive a simulation of another computer nicely addressed this drawback of generated workloads.

In these early days of trace analysis and trace driven simulation (TDS), nearly all programs were sequential programs. As shared memory multiprocessors became a commercial reality, architects and system designers again

93

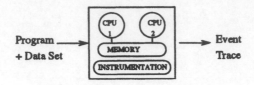

Figure 5.1: Collecting an Event Trace

used traces to predict the behavior of their designs. The subject programs for
these machines are multithreaded computations. However, the assumptions
that allowed sequential program traces to provide an arbitrarily detailed representation of the performance of a sequential program do not hold in the
shared memory processor environment due to the potential for nondeterminacy in multithreaded code.

A parallel program defines two or more threads/processes of execution
that can be executed at the same time, or in some constrained order, to
complete a computation. It is well-known that if two or more threads share
variables, it is usually necessary to synchronize the threads whenever they
read or write the variables; otherwise a consumer thread might read a different value from a variable on two different executions depending on whether
or not a producer thread wrote into the variable prior to, or after, the consumer's read operation. Nondeterminacy occurs in other guises in the shared
memory multiprocessor environment. For example, if two or more threads
are blocked on a lock variable in shared memory, then multiple executions of
the same program on the same data may result in different threads gaining
the lock in different executions. Another example is the potentially nondeterministic behavior of a multiprocessor's cache manager, allowing different
executions of the same program on the same data to cause different cache
lines to be loaded and unloaded.

Why is this such an important problem in considering traces and share
memory multiprocessors? Consider the specific code schema shown in Figure 5.2. If the child was the last process to assign a value to k prior to its
execution of the if-statement, it will **exec** prog1; but if the parent was the
last process to assign a value to k, the child will **exec** prog2. The trace
for the cases will be completely different, depending on the outcome of the
race on any particular execution. The presence of parallelism and nondeterminacy means there are at least two different traces for the execution of
this parallel program. In this case, depending on either trace as an accurate
representation of the workload is misleading.

If nondeterminacy is possible in a computing scenario, any particular
run of the program on a given data set may result in a different trace from

```
shared int k;
...
if(fork() == 0) {
    k = 1;
    if(k%2 == 0) execve(prog1, ...);
    else execve(prog2, ...);
}
k = 2;
...
wait(...);
```

Figure 5.2: A Nondeterministic Code Fragment

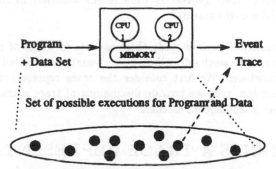

Figure 5.3: Traces of Nondeterministic Programs

another run. In Figure 5.3, we highlight this possibility by indicating that any specific execution corresponds to only one member of the set of all possible traces for the given program/data/architecture. This fundamental observation leads us to consider three important problems regarding the use of event traces in multiprocessors:

Trace Representativeness. Given the possibility of many different traces, is the particular trace obtained from an execution "representative" of the other possible traces for that program. Will an analysis of any particular trace result in the same conclusion as the analysis of any of the other traces? If the trace drives a simulation, will the results from the simulation be similar for any of the traces?

Trace Extrapolation. The rationale for TDS is perhaps even more significant in multiprocessor traces than in sequential traces, since speedup and data movement are so critical to the program's behavior. The technique is to use a trace obtained from one multiprocessor to define the program load on a parametrically-related multiprocessor. For example, one might trace the execution of a program on a 4-CPU configuration, then extrapolate the trace for a 16-CPU configuration of the same machine type. Is it possible to derive such a trace so that it is correct and representative? What is the computational cost for deriving accurate traces?

Trace Approximation. Sometimes analysts extrapolate an apparently reasonable trace even though the trace could not actually have occurred in *any* execution of the program. Such approximations may be computationally simple to derive, yet be as representative as an actual trace. How can one reason about trace generation time versus accuracy in the context of nondeterministic environments?

This paper describes work to study these aspects of traces of nondeterministic programs. The results described here summarize detailed formal work appearing elsewhere.[1] We first consider the trace representation problem in the next section, then we provide discussions of trace extrapolation and approximation in subsequent sections.

5.2. WHEN IS A TRACE REPRESENTATIVE?

Let us refer to the set of all possible traces for the execution of a nondeterministic program as Γ (see Figure 5.3). Each element, $\gamma \in \Gamma$, corresponds to one trace that could be derived from the execution. Given $\gamma_i, \gamma_j \in \Gamma$, it is possible to define a *metric*, $M(\gamma)$ and a corresponding *distance*, $D(\gamma_i, \gamma_j) = | M(\gamma_i) - M(\gamma_j) |$. For example, $M(\gamma)$ might be the number of event occurrences in the event trace, so $D(\gamma_i, \gamma_j) = 0$ if the two traces are the same length.

A metric partitions the set of traces into equivalence classes by grouping traces with the same metrics into the same class. If one selects a metric where all traces are in the same equivalence class, then it could be argued that on the basis of the metric, any trace from the class is representative of all the others since they all have the same metric. In the case there is more than one equivalence class, we can quantify the degree to which traces differ by computing the distance between members of the two classes. This provides us with a quantitative scheme for assessing the representativeness of a trace. The choice of a metric reflects the perspective used to consider two traces, and is a subjective exercise. Hence, the determination of whether or not a trace is representative is subjective.

How can one choose a metric that is meaningful? Presuming the program halts in every execution (on a particular data set), Γ is finite, though potentially very large. The size of the set is determined by the number of times a nondeterministic choice can be made during the execution. For example, in Figure 5.2, if k is the only shared variable between the two processes, and all other parts of the execution (the parent, prog1, and prog2) are deterministic, then $| \Gamma | = 2$. We call the point where any such nondeterministic choice is made — the statement branching on the value of k in the example — the *trace change point* (TCP). By restricting the way parallel programs can be created and terminated, and the language constructs within the program, it can be shown that for every TCP where two traces diverge, there is a corresponding point, called the *convergence point* (CP), in the two traces where they converge onto the same event pattern in the trace.[1] Two traces, γ_i and γ_j might be considered to be in the same equivalence class if they have the same number of TCP-CP pairs; a more refined measure is that not only do they have the same number, they also have the same set of TCP-CP pairs, though the computations between the TCPs and CPs might be different (i.e., each equivalence class may still contain many different traces).

The TCP structure of a trace is a fundamental characteristic related to the nondeterminism in the execution, and therefore a critical parameter for considering metrics that partition Γ into equivalence classes. Hence, a fundamental operation in determining the representativeness of a trace is to be able to find and analyze the context for TCP occurrences in a trace. We observe that every TCP event in a trace is caused by the execution of an identifiable TCP construct in the program. Identifying TCP occurrences in a trace requires the knowledge of which statements of a program are TCPs. Unfortunately, it can be demonstrated that the task of locating all TCPs in a parallel program, using static analysis of the program, is an NP-hard problem.[1] We have derived a data flow algorithm to compute a conservative estimate of the set of TCPs of a parallel program in $O(N \times E \times V^2)$, where N is the number of statements in the program, E is the number of edges in the program statement flowgraph, and V is the number of variable in the program. That is, the algorithm locates all actual TCP-CP pairs (along with spurious other locations).

The time taken to locate program TCPs is a lower bound on the complexity of structural metrics. It is still possible to choose a metric that is unreasonable to calculate.

5.3. TRACE EXTRAPOLATION

In trace driven extrapolation, a trace is collected on a *source machine*, then used to define a trace for execution on a parametrically related *target ma-*

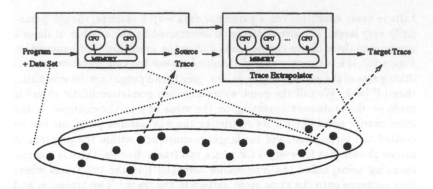

Figure 5.4: Trace Extrapolation

chine (see Figure 5.4). Since the target machine usually does not exist, the source trace must be translated from the source execution architecture to the target execution architecture by simulating selected parts of the target.

The source trace, γ_s, is a member of the set of possible executions, Γ_s, for the given parallel program and input data set executed on the source machine, but the process of moving the source trace to a new execution architecture may generate a new target trace, γ_t that is a member of the set Γ_t of traces of the program for the target machine, and may or may not be a member of Γ_s. This arises due to execution order distortion in the derived target trace, i.e. the path taken during the execution that produced the source trace may not be the path taken for the target trace. If an extrapolation method cannot be shown to always produce a trace in Γ_t, the determination of membership in Γ_t becomes an important problem for trace extrapolation.

Ideally, γ_s would be *reused* to produce the target trace, γ_t, i.e., the program was used to produce the source trace and it should not be necessary to *reexecute* the portions that contributed to the source trace. More precisely, *the problem of trace extrapolation is that of taking a trace on an existing execution architecture, given a program and data set, and adjusting the trace without the use of program reexecution to produce a trace that is a correct trace for a parametrically different execution architecture* (i.e., resulting in a trace that would have occurred had the program and data set been run on the hypothetical execution architecture). Determining whether or not it is possible to solve the trace extrapolation problem is the fundamental question of whether or not trace correctness can be guaranteed for simulations.

Holliday and Ellis present a technique for extrapolating parallel program traces.[2] The technique relies on static analysis of a parallel program to identify the points in the trace where it changes from one trace instance, γ_i, to another (i.e., in locating the TCPs in the trace). As we mentioned in Section 5.2, the problem of statically locating TCPs and CPs is an NP-hard problem.

Holliday and Ellis leave open the question of the complexity of their solution and of the trace extrapolation problem in general. We have demonstrated that this problem is in general NP-hard.[1] We have also attempted to concisely identify programs that produce executions sets for which trace extrapolation is tractable; this implies we have considered the *level of nondeterminism* of a parallel program. Various characterizations of program nondeterminism exist,[4,5] however these characterizations identify the nondeterministic behavior of a program based on the data values that the program computes. That is, a program is *data nondeterministic* if it has the possibility of producing more than one set of data values on a given input data set. This characterization is not sufficient for considering traces, since a data nondeterministic program may produce traces comprised of a single set of events. We call such programs *execution pattern deterministic* (or simply pattern deterministic). For pattern deterministic programs, trace extrapolation is a simple process. Whereas for pattern nondeterministic programs, those producing traces comprised of possibly different sets of events, trace extrapolation is complicated. The structure present in the set of executions from a parallel program, allows us to concisely identify the program as pattern deterministic or pattern nondeterministic.

Elsewhere,[1] we provide a formal model in which to represent nondeterminism and to study the trace extrapolation problem. The model again uses trace equivalence to group sets of structurally similar executions. Equivalences are classified as either *data and pattern deterministic, data nondeterministic and pattern deterministic,* or *pattern nondeterministic* with the following relationship:

Data and Pattern Det. $\subset$ Data Nondet. and Pattern Det. $\subset$ Pattern Nondet.

We introduced a family of equivalences for pattern nondeterministic programs. This family is intended to provide a means of differentiating nondeterministic program behavior. This measure coupled with the number of TCPs present in a program represents an indication of the size of a trace extrapolation problem instance.

Within the framework, equivalences that lead to sets of executions for which trace extrapolation is tractable can be identified. These equivalences identify program behaviors that lead to tractable or intractable trace extrapolation. Using this framework, it can be demonstrated that, for the smallest choice of equivalence in the family of pattern nondeterministic equivalences, the trace extrapolation problem is in general NP-hard. That is, there are no classes of pattern nondeterministic program for which trace extrapolation is tractable.

5.4. TRACE APPROXIMATION

Trace-driven simulation is the seminal example of the necessity for trace extrapolation. The target environment for a TDS may not exist and the trace used for simulation may be from a parametrically different environment. Trace driven simulations can be divided into two categories: *synchronous trace driven simulation* and *asynchronous trace driven simulation*. In synchronous TDS, the event ordering of the source trace is adopted in the target trace. Specifically, the behavior of the source environment with respect to timing (e.g., lock acquisition), is assumed to be correct for the target environment. While synchronous TDS is accurate for sequential programs running on uniprocessor environments, where extrapolation is unnecessary, Goldschmidt and Hennessy empirically show the technique to be inaccurate for multiprocessor computer systems.[4] That is, when conducting simulations of multiprocessor systems, it is not possible to ignore timing differences between host and target machines. In asynchronous TDS, the execution architecture of the simulated environment determines the event ordering for the target trace. Asynchronous TDS provides accurate simulation results for pattern deterministic programs. However, our work with trace extrapolation demonstrates that the use of asynchronous TDS in conjunction with pattern nondeterministic programs cannot be guaranteed to provide accurate simulation results. Because trace extrapolation for pattern nondeterministic programs is intractable, TDS cannot in general provide a correct trace.

In order to make asynchronous TDS tractable for pattern nondeterministic programs, a set of executions for which trace extrapolation is tractable must be considered. That is, the extrapolation process must be approximated. One difficulty that arises due to approximation is that extrapolation may yield an execution outside of the set of executions under consideration. We can demonstrate that the problem of determining membership within a set, derived from an equivalence class and given a single member of the set (i.e., the source trace), is undecidable for all but data deterministic sets.[1] This implies that the approximation is set of executions must be augmented with executions that are not in Γ — the set of *feasible* executions. It should be noted that the actual effect of using potentially incorrect traces for asynchronous TDS has not been resolved.[3]

Our intuitive model of trace approximation is shown in Figure 5.5: a source trace is generated on an existing execution architecture; this trace is then extrapolated to a new trace based on the behavior of the execution

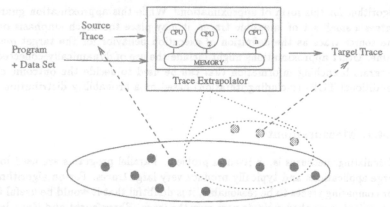

Set of executions for a program and data set including some approximations

Figure 5.5: Approximate Trace Extrapolation

architecture under study. In order to achieve tractable extrapolation, we employ approximation techniques that are based on a metric that yields a tractable extrapolation process, though we may generate target traces that are not feasible. Thus, approximation potentially creates $\gamma_k \notin \Gamma$ but for which γ_k has a measure, $M(\gamma_k)$ such that it is representative of some $\gamma_i \in \Gamma$. The question then becomes one of whether or not the use of γ_k in a TDS will provide adequate simulation results (i.e., will be sufficiently accurate). That is, is the approximation metric sufficient for generating accurate traces.

Approximation complicates the process of defining a calculable and representative metric by further restricting the metric such that it can be used to define a set of executions for which trace extrapolation is tractable. There are a spectrum of possible choices of metrics for approximation: from the restrictive choice relating traces that execute the same events in the same order (i.e., data determinism or synchronous TDS), to the optimal choice of relating traces from the same program, data set, and execution architecture (i.e., the general trace extrapolation problem). One reasonable choice is the metric that equivalences a source execution together with all executions, feasible or infeasible, that execute the same set of TCPs with the same outcome. Here, trace extrapolation is tractable (since there are no longer TCP outcomes to decide). Consider again the example in Figure 5.2. Assume that in some feasible source trace, executing the conditional TCP results in an **exec** of prog1. Then this metric would group the source trace with the infeasible trace that contains an **exec** of prog1 even though the parent is the last process to assign a value to k. In other work we presented an algorithm for this form of approximation.[1] While this approximation guarantees a small set of infeasible traces, it may place too much emphasis on the source trace as the definition of execution behavior for the target machine. Other approximations might include the use of dynamically collected program branching information that can be used to decide the outcome of conditional TCPs (including iteration) based on a probability distribution.

5.4.1. Measurement

Calculating measures is, in itself, a problem. Parallel programs are used for large applications and typically produce very large traces. For an algorithm for comparing traces to be reasonable, it is doubtful that it would be useful if it required more than a single pass over the trace. Ehrenfeucht and Haussler have shown it is possible to compare sequences in linear time,[6] but these algorithms will be unreasonable for large sequences due to the necessity of reversing an input sequence. We make the observation that executions exhibit periods of divergence and convergence. Specifically, that executions can diverge at TCPs, but will converge at the CP corresponding to the divergent TCP. This observation allows us to take advantage of the program structure to analyze the trace; by correlating the program structure with the trace information, it can be shown that a correspondence between two executions can be computed on-the-fly (i.e., in linear time and constant space).

While there are many opinions about what makes a trace accurate, currently, there are few characterizations of the accuracy of traces used in simulations.[2,4,5,7] The problem of determining the accuracy of an approximation is one of measuring how far, by some metric, the target trace is from the actual target trace (i.e., the trace that would have occurred had the program been executed in the target environment unobserved). This problem is trivial for pattern deterministic programs since the target trace is the actual target trace. For pattern nondeterministic programs, let S be the source trace, T be the derived trace, and A be the *correct* actual target execution. The distance between the source and actual target execution is $d_{(s \to a)}$, the distance between the source and target execution is d_f, and the distance between the target execution and the actual target execution is $d_{(t \to a)}$. Because the actual target trace is unknown, the distances $d_{(s \to a)}$ and $d_{(t \to a)}$ are not exactly calculable, while the distance d_f is exactly calculable. In Eckert's thesis, she demonstrated the fundamental limitations of placing a bound on the $d_{(s \to a)}$ and $d_{(t \to a)}$ distances using the source and target traces and information gained during the extrapolation process.[1] The usefulness and accuracy of these bounds remains to be demonstrated.

6. CONCLUSION

Issues surrounding the validity of trace driven simulations of shared memory multiprocessors, have led us to consider three important problems regarding event traces. Whether or not a trace is representative of the behavior of a program is a subjective determination. However, the notion of a metric and distance between traces provides a quantitative scheme for accessing trace representativeness. For deterministic trace generation systems that produce a single trace for a given program and execution architecture,[6,9,10] the issue of what execution behavior is representative is critical.

Trace driven simulation is the seminal example of the necessity for trace extrapolation. Extrapolation it is also critical for trace collection and perturbation analysis. Using a framework based on the distinction of pattern determinism and pattern nondeterminism, we have demonstrated that the trace extrapolation problem is in general NP-hard. For programs that are data deterministic (e.g., programs containing post/wait synchronization with no clear), trace extrapolation can accomplished in linear time and space. For programs that are pattern deterministic but potentially data nondeterministic (e.g., programs containing DOALLs and using floating point arithmetic), trace extrapolation is also a linear time and space process. However, for pattern nondeterministic programs, those using arbitrary synchronization (e.g., spin locks), trace extrapolation is intractable. There are many examples of intractable problems that, in practice, pose no serious computational problems. However given the potential size of parallel program traces and the potential for a parallel program to produce a set of executions of significant cardinality, it seems unlikely that current algorithms for trace extrapolation will perform well in practice.

Synchronous TDS is inaccurate for simulating the behavior of shared memory multiprocessors. For pattern deterministic programs, asynchronous TDS is accurate. However for programs exhibiting pattern nondeterminism, trace extrapolation must occur and hence, asynchronous TDS cannot guarantee a simulation based on a correct trace. For this reason, approximations to the trace extrapolation process must be considered.

One way to deal with approximations would be to supply a measure of the accuracy of the approximation. Ideally, this measure would be the distance between the target trace and the actual target trace, (though the target machine may not exist, meaning it may not be possible to obtain the actual target trace). The structure induced on a trace by a program makes it possible to make a correspondence between two traces from the same program that will identify periods of convergence and divergence on-the-fly (i.e., in linear time and constant space). Therefore, the complexity of actually computing a distance between two traces is dominated by the time to compute the actual distance. Although it is possible to measure distances, there are fundamental limitations on what can be gleaned from a source trace, target trace, and the extrapolation process.

Asynchronous TDS based on approximations is currently being used for the simulation of shared memory multiprocessors. Yet currently it is not known if infeasible traces produce accurate simulation results and it is also not known if there is a correlation between the degree of inaccuracy present in a trace and the accuracy of simulation results. Until the validity of trace driven simulation using approximation techniques has been demonstrated, results based on such simulations of shared memory multiprocessors must be considered suspect for programs exhibiting pattern nondeterminism.

REFERENCES

1. ZULAH K. F. ECKERT, Trace Extrapolation for Parallel Programs on Shared Memory Multiprocessors, Ph.D. thesis, University of Colorado, Department of Computer Science, 1995.

2. MARK A. HOLLIDAY and CARLA S. ELLIS, "Accuracy of Memory Reference Traces of Parallel Computations in Trace-Driven Simulation," IEEE Transactions on Parallel and Distributed Systems, 3, 1, pp. 97–109 (1992).

3. P. A. EMRATH and D. A. PADUA, "Automatic Detection of Nondeterminacy in Parallel Programs," in Proc. Workshop on Parallel and Distributed Debugging (ACM Press, New York, 1991), 89–99.

4. STEPHEN R. GOLDSCHMIDT and JOHN L. HENNESSY, "The Accuracy of Trace-Driven Simulations of Multiprocessors," in Proceedings ACM Sigmetrics Conf. on Measurement and Modeling of Comptuer Systems, (ACM Press, New York, 1991), 146–157.

5. P. BITAR, "A Critique of Trace-Driven Simulation for Shared-Memory Multiprocessors," Cache and Interconnection Architecture, 1990, 27–52.

6. A. EHRENFEUCHT and D. HAUSSLER, "A New Distance Metric on Strings Computable in Linear Time," Discrete Applied Mathematics, 20, pp. 191–203 (1988).

7. E.J. KOLDINGER, S. J. EGGERS, and H. M. LEVY, "On the Validity of Trace-Driven Simulation for Multiprocessors," in Proc. of the 18th Symposium on Computer Architecture, (IEEE Computer Society Press, 1991).

CHAPTER 6
MEMORY MANAGEMENT AND SPEEDUP
ISSUES IN PARALLEL SIMULATION

Mohamed S. Ben Romdhane and Vijay K. Madisetti

6.1. Introduction

Recent advances in technology, especially in communication networks and computer systems, have stimulated the search for new modeling and analyzing tools. Except for very few particular cases, these systems are analytically intractable, and numerically prohibitive to evaluate. One often has to resort to simulation as the only available methodology. Unfortunately, conventional sequential simulations of complex discrete event systems are exceedingly slow[1]. One alternative is to partition the simulation problem and execute the parts in parallel. However, because of order-preserving constraints, conventional parallelization techniques provide little benefit[27]. An entirely new approach to simulation for multiprocessors called *parallel simulation* is required. Parallel simulation has completely different storage management problems from conventional simulation. An excellent overview of parallel simulation is found in[1]. Memory requirements of parallel simulation are highly dependent upon the synchronization approach and the semantics of the application. In order to assure a stable parallel simulation implementation, some storage management protocols have to be added to deal with unstable memory scenarios. An efficient storage management protocol is required in order to end up with interesting speedups. In this chapter, we first introduce the parallel simulation approach, then we

105

discuss memory issues in this context. Section 6.4 describes efficiency considerations and reported speedups. Finally, Section 6.5 presents our conclusions.

6.2. Parallel Simulation

An obvious means of obtaining a faster simulation is to dedicate more resources to it. Parallel simulation, also referred to as Parallel Discrete Event Simulation (PDES), attempts to speed up the simulation by executing a single simulation program on a parallel computer. The mapping of a simulation program on a parallel computer is achieved by decomposing the simulation application into a set of concurrently executing processes. The greatest opportunity for parallelism arises from processing events concurrently on different processors. However, this concurrent execution of events at different points in simulated time, introduces interesting synchronization problems that have to be resolved properly in order to allow any speedup. The difficulty of PDES is caused by the sequencing constraints that dictate the order in which computations must be executed relative to each other[1]. Sequential simulators typically utilize three data structures:

- State variables describing the state of the system.

- Event list that contains timestamped scheduled events, but not yet processed.

- Global clock describing the up to date simulation time.

The global structure of a sequential simulator is described by Figure 6.1. It is crucial for the simulation that one always selects the smallest timestamp event (E_{min}), from the event list, as the one to be processed next. i.e., consider the two following events:

1. E_1: Bus departure from station is scheduled at 10:00 am.

2. E_2: Passenger X's arrival to the station at 9:50 am.

If E_2 is processed before E_1, passenger X can ride the bus, and the simulation is correct. However, if E_1 is processed before E_2, passenger X will not be able to ride the bus even if he has arrived before its departure (imagine X is the driver of the bus, in this case the bus would

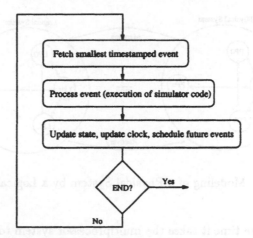

Figure 6.1: Global structure of a sequential simulator.

depart without its driver!). It is clear that such situations, where the future affects the past, are erroneous. These kind of errors are called *causality errors*.

6.2.1. Parallelization of a Simulation Program

The mapping of a simulation program into a parallel computer should take into account two major points:

- Correctness,
- Efficiency.

Correctness means that the parallel simulation should give the same results as those of the correct sequential simulation. *Efficiency* means that the total time for execution of the parallel simulation is less than that for the execution of a functionally equivalent sequential simulation. Another definition of efficiency is given by:

$$Efficiency = \frac{Speedup}{Total\ number\ of\ processors\ used\ by\ the\ simulation}$$

and measures the effective utilization of the processors. *Speedup* is defined as the time it takes a single processor to perform a simulation,

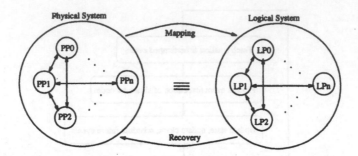

Figure 6.2: Modeling of a Physical System by a Logical System.

divided by the time it takes the multiprocessor system to perform the same simulation. Ideally the speedup is N (the total number of processors used by the simulation), and the efficiency is 1. In this definition of speedup, the single processor should have the same CPU capabilities as one of the processors in the parallel computer system. We are concerned with parallelization of the simulation of asynchronous systems, where events are not synchronized by a global clock, but rather occur at irregular time intervals. Upon the occurrence of an event, the simulation model jumps from one state to another. Because of the dynamic nature of PDES problems, general parallelization solutions (i.e., vectorization techniques using supercomputer hardware) provide little benefit[27]. The *Physical System* (PS) to be modeled is composed of a set of interacting *Physical Processes* (PP). Every PP is modeled by a *Logical Process* (LP), and interactions between PPs are modeled by exchange of timestamped messages between LPs, as shown in Figure 6.2. The timestamp is the simulated message arrival time at the receiving process. Every LP contains the state corresponding to the PP it models and a clock denoting how far the process has progressed.

In order to insure correctness of the parallel simulation, one has to make sure that no causality errors occur, that is LPs should adhere to the local causality constraint[1].

Local causality constraint[1]- A discrete event simulation consisting of logical processes (LPs) that interact exclusively by exchanging timestamped messages obeys the local causality constraint if and only if each LP processes events in non-decreasing timestamp order.

6.2.2. Parallel Simulation Approaches

Depending on how synchronization is enforced, PDES mechanisms fall into two categories : *Conservative* and *Optimistic*.

6.2.2.1. Conservative Mechanisms

Conservative mechanisms developed by Chandy and Misra[5,6], strictly avoid causality errors. Only safe events are processed. A logical process decides that event E is safe if it is assured that no other event with a smaller timestamp than that of E will ever be received. When a LP runs out of safe events (i.e., one of its incoming links is empty), it must block. As a consequence, if appropriate precautions are not taken, deadlock situation are likely to happen; Figure 6.3[1] shows such situation. Here, each process is waiting on the incoming link containing the smallest link clock value because the corresponding queue is empty. All three processes are blocked, even though there are event messages in other queues that are to be processed. Two different approaches were proposed to address the deadlock problem in conservative simulation. The first, *deadlock avoidance*[5,10], prevents deadlock occurrence through the use of synchronization messages, usually called *null messages*. These messages, however, do not correspond to any activity in the physical system. The second approach, *deadlock detection and recovery*[6], addresses the problem after occurrence. A separate mechanism is used to detect when the simulation is deadlocked; if a deadlock is detected, still another mechanism is invoked to break the deadlock. A process that can predict at simulated time T all events that it will generate up to simulated time $T + L$ is said to have lookahead L. Lookahead is used in the deadlock avoidance approach to determine the timestamps that are assigned to null messages.

6.2.2.2. Optimistic Mechanisms

Optimistic mechanisms are based on the *Virtual Time Paradigm* introduced by Jefferson in[11,12]. Virtual time paradigm is a way of organizing distributed systems by imposing on them a temporal coordinate used to measure computational progress and to define synchronization. Virtual Time (VT) may or may not have a connection with real time. However, it provides a flexible abstraction of real time in the sense that virtual clocks can jump backward. Every LP in the

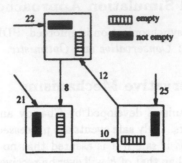

Figure 6.3: [1]A deadlock situation.

virtual time system maintains its own Local Virtual Time (LVT). LPs always process input messages whenever they are available, under the implicit assumption that the message sequence does not violate causality. While all virtual clocks tend to go forward toward higher virtual times, they occasionally jump backward. These backward jumps, usually called *Rollbacks*, occur after a causality violation. In this case the LP is returned to a state immediately prior to the violation, and execution resumes. The most well known optimistic protocol based on the virtual time paradigm is *Time Warp*[11]. However, another promising optimistic protocol, known as *Probabilistic Synchronization* was introduced by Madisetti[16,17]. This protocol differs from Time Warp in the sense that processors are kept in a loosely synchronous scheme through additional synchronization means described by probabilistic synchronization calls. Recently, Chandy and Sherman[7] also introduced a new optimistic approach known as *Space-Time Simulation*, which attempts to extract parallelism from both time and space.

Time Warp

The representation of a Time Warp process is composed of:

1. Process identity.

2. Local Virtual Time (LVT), denoting how far the LP has progressed in virtual time. It is the timestamp of the event being processed.

3. Actual state of the LP, which is in general the entire data space of the process.

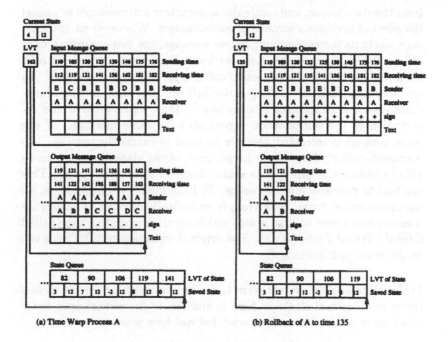

(a) Time Warp Process A   (b) Rollback of A to time 135

Figure 6.4: [1]LP structure and rollback.

4. State queue, containing the history of state vectors down to the Global Virtual Time (GVT, defined below).

5. Input queue, containing history (down to GVT), present, and future events sorted in order of virtual received time.

6. Output queue, containing antimessages corresponding to the messages recently sent by the process. In case of rollback, antimessages are used to cancel the previously erroneous sent messages.

Figure 6.4(a)[11] shows the structure of a LP in Time Warp, while Figure 6.4(b)[11] shows the rollback of the same LP to virtual time 135. The event that causes the rollback is called a *straggler*. Recovery is accomplished by undoing the effect of events that have timestamps larger than that of the straggler. Whenever a process sends a message (positive), it keeps a copy (negative) in its output queue. These two copies are differentiated by a flag. A rolling back processor restores an old correct state from the state queue, and sends the appropriate antimessages to cancel the effect of previously erroneous sent messages. Whenever an antimessage meets its corresponding positive message, the two annihilate each other. When antimessages are sent immediately after a process rolls back, the cancellation approach is called *aggressive cancelation*[11]. In an other approach called *lazy cancellation*[8], processes do not immediately send antimessages after rollback; they wait to see if the reexecution of the computation regenerates the same message. If the same message is recreated, there is no need to cancel the message. As discussed earlier, Time Warp keeps track of old states in order to be able to restore the appropriate state whenever a rollback occurs. This can lead to excessive memory usage. To limit memory consumption, an algorithm called *fossil collection*, is periodically invoked to deallocate memory being used to store states which occurred prior to a time called *Global Virtual Time* (GVT). The original definition of GVT is given by Jefferson and Sowizral[12].

Definition 1:*GVT at real time t, is the minimum of (i) all virtual times in all virtual clocks at time t, and (ii) of the virtual send times of all messages that have been sent, but not been processed at time t.*

GVT can be viewed as the virtual clock for the system as a whole, it measures the progress of the system.

Probabilistic Synchronization

In[17,16], Madisetti, Walrand, and Messerschmitt, and Madisetti and Hardaker described the theory of Probabilistic Synchronization through what they called *Self-Synchronizing Concurrent Computing Systems* (SESYCCS). In this type of SESYCCS, the deterministic synchronization environment is replaced by one in which the algorithms are themselves stochastic. In addition to the usual communications imposed on the computation by the dynamics of the algorithm, each processor i at some point in time n, decides whether or not it should communicate some "clock" information $C_i(n)$ to another processor j. In the following we discuss the Probabilistic Synchronization as was implemented in the MIMDIX Operating System for Simulation[18]. In this context, processes are synchronized by a special MIMDIX process called *genie*. The genie continually updates the GVT of the system and periodically (in the probabilistic sense) sends a broadcast synchronization (broadcast

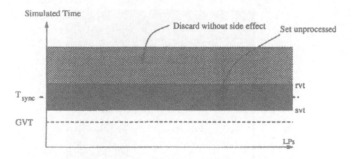

Figure 6.5: Resynchronization in Probabilistic Synchronization.

SYNC) to the system. The frequency of the broadcast SYNC is determined by a delay time (δ), and a probability of synchronization (p_s). In a tightly coupled system, the SYNC is timestamped with a value that is slightly larger than GVT. The broadcast SYNC mechanism is a shared memory variable. When set by the genie process, it signifies that a broadcast SYNC has been required. The logical processes, if they find out that the SYNC is set, execute the resynchronization routine. Fossil collection may preempt the resynchronization. The resynchronization routine discards all positive and negative messages with a timestamp greater than the SYNC timestamp, possibly rolling back LPs without sending antimessages. Already processed events, with send timestamps less than or equal to the SYNC timestamp (T_{sync}) and receive timestamps greater than T_{sync} are marked unprocessed. Figure 6.5 illustrates the resynchronization operation.

Space-Time Approach

Most simulation algorithms and performance studies have only exploited spatial decomposition of the model to be simulated. Chandy and Sherman[7] proposed a unifying framework called *Space − Time*. A new algorithm suggested by this framework was also proposed. The idea of the space-time approach is to exploit the temporal decomposition of the model as well. A discrete event simulation can be viewed as a two-dimensional space-time graph where one dimension describes the state variables used in the simulation, and the second dimension is simulated time. In order for the simulator to characterize the behavior of the physical system, it must determine the value of the state

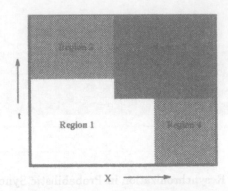

Figure 6.6: Space-Time decomposition.

variables over simulated time. In the framework proposed by Chandy and Misra, the space-time graph is partitioned into regions, with one process assigned to each region (see Figure 6.6). Each process computes the behavior of the model in the part assigned to it (we also say the process is "filling in" its region) and communicates its results to its neighbors at arbitrary times. A process recomputes the behavior in its region if it has new information about the behaviors in other neighboring regions. The algorithm declares completion and returns results at a fixed point (i.e., when recomputation always gives the same result). The convergence detection algorithm employs a special process called the detector. Further, there are communication channels from each process to the detector.

6.3. Memory Issues

Let us assume that we want to run a parallel simulation on a multiprocessor system having a huge amount of memory available for the simulator. The question that one may ask is: *does the multiprocessor system have enough memory to handle the simulation up to completion?*. There is no obvious answer to such question unless the simulation disposes of some kind of mechanism that can assure the survival of the simulation when it faces memory shortage problems. such mechanisms are generally referred to as memory management algorithms, and are generally implemented in optimistic systems (i.e., Time Warp). Optimistic methods have completely different storage management problems and opportunities compared to conservative methods. The presence of roll-back as a synchronization tool allows much greater flexibility in the management of buffers in optimistic systems than is possible in conservative methods. This is because in conservative systems the act of accepting a message for buffering at the receiver's end is an act of commitment. Since processes are not allowed to roll back, a message can not be regenerated by its sender. Hence conservative mechanisms have far less flexibility in buffer management than optimistic mechanisms do. Jefferson[13] showed that, contrary to published claims by Chandy in[6], asynchronous conservative algorithms are not optimal; they cannot necessarily execute a simulation in the same amount of space as sequential execution. In fact, Jefferson showed that a simulation requiring space $n + k$ when executed sequentially, might require $O(nk)$ space when executed on n processors by Chandy-Misra algorithm.

6.3.1. Memory Management Algorithms

This section discusses memory management algorithms for optimistic parallel simulation. Such algorithms are classified into two categories: Storage Management of the *Past*, and Storage Management of the *Future*.

6.3.1.1. Storage Management of the Past

Most Time Warp descriptions and implementations simply rely on *fossil collection* as the only storage management mechanism. It is also called garbage collection because it recovers memory whose contents are so old that they cannot have any future effect on the computation. Fossil collection implements irrevocable actions that rely on the computation of the GVT. Allocated space for committed events (with timestamps earlier than GVT) along with their corresponding state vectors and output messages can be reclaimed.

6.3.1.2. Storage Management of the Future

Fossil collection (FC) is insufficient for stable memory management in optimistic mechanisms. Memory freed up by FC may not be sufficient for the simulation to resume. Other problems known as *flow control* problems may happen between LPs when producers are generating messages at higher rate than consumers can handle (run out of buffer space to hold the unprocessed messages). Mechanisms that address these kind of problems are called storage management of the future, because they reclaim resources used to hold uncommitted events (and state vectors). This is made possible because in optimistic mechanisms, uncommitted events can be regenerated if the relative producers are appropriately rolled back. Flow control algorithms using discarding and resending of messages (relying on the acknowledgment techniques) can not be used for optimistic systems, because uncommitted events can never be safely acknowledged (since LPs can rollback), and message routing and queuing is not order preserving.

Message Sendback

The basic idea of message sendback is that the process running out of memory returns previously received input messages with send time larger than GVT to their original senders to make room for new arriving messages[1]. Gafni[8] proposed a generalization of message sendback in the sense that the message to be sent back can be either an input message, a process state, or an output message. The element chosen to be sent back is the one with the largest send time, because it is the least likely to affect the progress of the object. Also because sending it back to its original sender slows down the fastest process in the system by rolling it back. In message sendback, GVT at time t is defined as the minimum of the values of all local clocks at time t and the send time of all messages in transit at time t. In this case when a process receives a sent back message with timestamp equal to GVT, message sendback may announce memory exhaustion.

Cancelback

Cancelback mechanism, recently introduced by Jefferson[13], is a generalization of Gafni's algorithm on sendback. It allows a process to manage objects in the whole system. If fossil collection cannot reclaim enough free memory to satisfy the request of process p, then some storage object of p has to be removed to make more room. Jefferson showed that by using cancelback, Time Warp can always execute until normal termination. The key to this result is that he assumes that message buffers are allocated from a global shared memory. As a result one can always find an item (input message, output message or state vector) in any process whose send virtual time is greater than GVT which can be canceled and reproduced later. Canceling this item makes room for the simulation to proceed. If the item v to be canceled is an input message, it is removed from the input queue and sent in the reverse direction back to its original sender's output queue, where it will possibly invoke a rollback. If v is an output message, it is removed from the output queue, and is sent in the forward direction to the receiver where it will usually annihilate with its antimessage and possibly cause a rollback. Finally if v is a state vector, its cancellation means that it should be deleted and the process rolled back to the sent virtual time of the state (the send virtual time of the state is its saving time; it is also equal to its receive virtual time). Jefferson assumes instantaneous GVT availability.

Artificial Rollback

Lin and Preiss[24] proposed a memory management algorithm called *artificial rollback*. The idea of this algorithm is that without receiving a straggler, a process p may rollback its computation to a time earlier than its LVT. Artificial rollback is optimal (i.e., simulation can always complete if at least the same amount of memory as used by the sequential simulation is allocated for the parallel simulation) if implemented in a shared memory architecture. When fossil collection fails in producing enough storage, artificial rollback is invoked and the process with the latest local clock is rolled back to the second local clock. In several aspects, artificial rollback is equivalent to cancelback. In fact, the cancellation of an input message is equivalent to an artificial rollback of the sender of that message, and the cancellation of a process state (or an output message) in process p is equivalent to an artificial rollback of the same process.

Probabilistic Synchronization as a Memory Management Algorithm

In optimistic mechanisms optimizing both space performance and time performance is very difficult. However, the stochastic nature of *Probabilistic Synchronization* (PS) introduced by Madisetti *et al*[16,17] attempts to do so. In fact, PS retains the positive features of both conservative and optimistic synchronization mechanisms to a large measure, since it keeps processors in a loosely synchronous framework through the use of random synchronization messages without blocking them. On the other hand re-synchronization as discussed earlier frees up memory from the tail of the event list by deleting events with timestamps larger than the timestamp of the re-synchronization message (T_{sync})(see Figure 6.5). Committed events are still garbage collected by the FC mechanism.

6.3.2. Distributed versus Shared Memory

While distributed memory is desirable from the stand-point of generality and scalability, communication overhead can affect the performance of parallel simulation. Shared memory on the other hand can reduce the synchronization overhead incurred by process interaction by making message passing very fast. Consider the partitioning problem in parallel simulation, which consists of 2 phases:

- *Phase* 1: map the most communicating LPs on the same processor.

- *Phase* 2: execute load balancing between LPs within the same processor.

In shared memory, the partitioning problem reduces to the execution of *Phase* 2, which can still be eliminated by adapting some kind of global scheduling policy (i.e., LPs can run on any processor). The benefit from shared memory for conservative approaches can be considerable[38]. In fact, deadlock avoidance (which can be very expensive when the degree of branching in the communication graph is high) can be implemented at minimal cost. i.e., the sending of null messages can be implied by the modification of certain shared variables. When deadlock can not be avoided (i.e., the benchmark contains cycles with zero processing time), detection and recovery have to be invoked. In general, in order to minimize deadlock detection cost, the relative algorithm runs infrequently. Shared memory can still improve deadlock detection. i.e, idle processors check an appropriate shared variable indicating the up to date number of idle processors. We can see that all shared memory benefits for conservative mechanisms evolve around avoiding explicit message exchange through the the use of shared variables. In optimistic mechanisms (i.e., Time Warp), however, shared memory benefits greatly from memory management algorithms. Indeed, algorithms like Cancelback and Artificial Rollback are optimal only if implemented in a shared memory environment. This is of great importance since when conservative approaches fail because of memory shortage problems, Time Warp with an optimal memory management algorithm is assured to succeed. Shared memory, as will be discussed later, can also reduce error propagation in Time Warp (direct cancelation, Fujimoto[2]).

6.4. Speedup Issues

This section discusses the efficiency considerations and reported performances.

6.4.1. Efficiency Considerations

In addition to previous considerations taken to develop a correct implementation of the parallel simulation, certain pitfalls unique to optimistic execution must be avoided in order to end up with an efficient implementation of the simulator.

6.4.1.1. Cancelation Strategies

As discussed earlier, after the occurrence of a rollback, cancelation can be either aggressive or lazy. Depending on the application, lazy cancelation may either improve or degrade performance. However, an interesting property of lazy cancelation is that computations with incorrect (or partially incorrect) input may still generate correct results (i.e., logical operations). Therefore, one may execute some computations prematurely. As a result, lazy cancelation can allow the computation to be executed in a time that is less than that required by the critical path execution[28,30]. On the other hand, when lazy cancelation is adopted, erroneous computations may spread further than they would under aggressive cancelation, resulting in a degradation in performance. Also, output messages occupy buffer space for a longer time then they do under aggressive cancelation. One might choose then to implement lazy cancelation if the simulator has efficient error cancelation algorithm (i.e., Wolf[19]), and an optimal memory management algorithm.

6.4.1.2. State Saving Cost

In order to rollback, a process should be able to restore the state vector corresponding to the last correct computation. Therefore, each LP must periodically save state (often referred to as check-pointing). However, state saving has two negative effects on both time and space. Memory is consumed and processing time is needed for state check-pointing. State saving overhead can seriously degrade the performance of optimistic mechanisms. Fujimoto[2] reports that the performance of queuing network simulations under Time Warp was **cut in half** when the size of the process's state was increased from 100 to 2000 bytes. One approach to reducing the state saving overhead is to reduce its frequency. However, in this case, the process might have to rollback to an earlier checkpoint-ed state and recompute the desired state from an earlier state. In this case, the penalty associated with state construction depends on the frequency of check-pointing. When the computation granularity to process an event is significantly large as compared to state saving overhead, Time Warp can still be effective. Another systematic solution to state saving cost is to use hardware support[4].

6.4.1.3. Process Scheduling Algorithms

When more than one process coexists within the same processor, some scheduling policy should be adopted in order to select the LP that should be executed next. *Round Robin* (RR) scheduling algorithm is simple to implement. However, it fails to take into account the temporal interaction of events and LPs. Minimum Virtual Time (MVT)[23] scheduling algorithm gives high priority to the LP with the smallest LVT. It then attempts to balance the virtual clocks within one processor by advancing the smallest LVT on each process. This approach helps the GVT to advance better, which results in a steady progress of the simulation. Fossil collection can also be invoked more frequently in this case. Another scheduling policy approach, the Minimum Message Timestamp (MMT)[23], reduces the number of rollbacks by giving high priority to the LP with input message having the smallest timestamp.

6.4.1.4. Error Propagation

Error can propagate throughout the system at a rapid rate. Such a situation may arise when little computation is required to process incorrect events, and the amount of parallelism in the application is less than the number of processors (there is always CPU resources to process erroneous computations as soon as they arrive). One approach to remedy this problem is to give anti-messages higher priority then positive ones. One can also give higher priority to the computation associated with the smallest timestamp. However, these remedies might not work well for applications with small parallelism as compared to the number of processors. Madisetti, Warland and Messerschmitt[19] propose a mechanism called *Wolf*. Whenever node i detects a straggler, a Wolf call freezes the computation within a "sphere of influence" of the straggler. A broadcast cancelation is then issued by node i to cancel the erroneous computation. The effectiveness of the Wolf algorithm is highly dependent upon its ability to compute a precise sphere of influence.

Another approach to prevent incorrect computations from propagating too far ahead into the simulated time future is the *Moving Time Window* (MTW)[29]. Only events within the time window are eligible for execution. However, it is not clear how the size of the window should be determined. The major disadvantage of this approach is that such window can block correct computations because it can not distinguish them from erroneous ones.

Fujimoto proposes the *direct cancellation* algorithm[2]. In order to limit the spread of erroneous computation, direct cancelation has to be implemented in a shared memory architecture. Whenever a process schedules an event for another process, it keeps track of a pointer to that event. This pointer is used if it is later decided that the previously scheduled event should be canceled.

6.4.2. Performance

Many recent implementations have been successful in extracting considerable parallelism from real world simulation problems. Some of them adopted conservative approaches, but the majority used optimistic mechanisms. In conservative approaches, lookahead appears to be essential to obtain significant speedups. Timestamps increment is the most obvious lookahead that the simulation can exploit. i.e., in a communication network simulation the timestamp increment function represents the amount of queuing delay incurred by messages while they wait for communication channels to become free.

Kleinrock and Felderman[22] proposed a model for a two processor

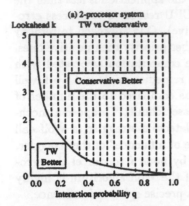
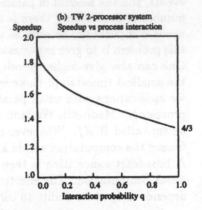

Figure 6.7: (a) [22]Using lookahead Conservative approach can outperform Time Warp. (b) [21]High process interaction affects the speedup of Time Warp.

system using conservative synchronization. The model assumes that processes have k-step full lookahead (i.e., a process is able to give the other process the content of any message up to k virtual time units in the future, by means of null messages). They showed that lookahead is extremely useful when processors are nearly balanced in processing speed. In this case, as illustrated by Figure 6.7(a)[22], conservative approach can easily outperform Time Warp (the comparison does not exploit lookahead for Time Warp) for small lookahead and high process interaction; however, when process interaction is small, conservative method can outperform Time Warp only when the lookahead is large. The Time Warp model used for comparison is their own two-processor model introduced in[21], where they first present a Discrete time Discrete state (DD) TW model, then they expand it to Continuous time Discrete state (CD), Discrete time Continuous state (DC), and Continuous time Continuous state (CC). In this two-processor TW model, they show how interaction between processes affects the speedup as illustrated by Figure 6.7(b)[21].

Quoting Fujimoto[1], all existing models for Time Warp require a number of simplifying assumptions. Rollback cost and state saving overhead are critical assumptions. Completely different results can be obtained depending on the assumptions made. Some researchers (Mitra and Mitriani[32], Lavenberg and Muntz[31]) assume that rollback cost is proportional to the rollback distance. Others (Lin and Lasowska[26]) attribute no cost to rollback. Madisetti, Walrand, and Messerschmitt[20], Lipton and Mizell[33], and Nicol[34] associate a fixed cost with rollback. One may find it reasonable to assume that the rollback cost is proportional to the length of the rollback since the principle overhead incurred is the time required to send anti-messages. Fujimoto[1] argues that since rollback only requires sending pre-existing anti-messages and that sending a positive message takes much more time (buffer allocation, data transfer), thus in practice the rollback cost will be much smaller than that of forward computation. As a consequence rollback cost can be assumed zero. This can be true when the length of the rollback is limited. However, in situation such that of cascading rollbacks[37], this assumption may seem very optimistic. Lin and Lasowska have shown that if no cost is associated with rollback and state saving, and if incorrect computations never rollback correct ones, then Time Warp using aggressive cancelation will produce optimal performance. Nicol[34] argues that this is intuitive and shows how assumptions can influence the results. In fact, because correct computations are on the critical path, if Time Warp is able to follow the critical path without cost, then its performance should be optimal.

Madisetti et al.[17] have developed a performance model that estimates the rate at which simulation time advances under their own strategy (Probabilistic Synchronization). Processors are classified into one of two sets: fast or slow. However, this distinction is not restrictive, a processor belonging to one of the two sets can change its dynamics and join the other set. As illustrated by Figure 6.8, in order to reduce the possibility of a cascade of resynchronizations after a slow processor communicates with a faster processor, all the fast processors are resynchronized.

Researchers at JPL have reported interesting results in simulations of communication networks (Warpnet)[35]. The version of Time Warp used for this simulation supports lazy cancellation and Message Sendback[8]; however, it does not support dynamic creation and destruction of objects. GVT was calculated every six seconds. Preliminary benchmarking of Warpnet, using a 32-node JPL/Caltech Mark III hypercube indicates a maximum speedup of 16.2 on 32 nodes. However, these performances were obtained at considerable memory expense. Warpnet requires each computer object to retain a copy of the entire network in

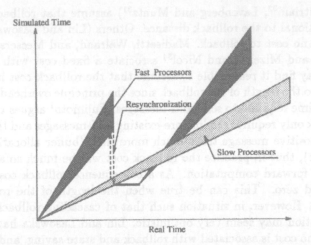

Figure 6.8: Performance Model: After a slow processor communicates with a fast processor, all the fast processors are resynchronized.

memory for routing purposes. JPL researchers have also reported interesting speedups of discrete-event combat simulation (STB88)[36] on two architectures: the Caltech/JPL Mark III hypercube and the BBN Butterfly. Despite the strong irregularities of battlefield scenarios, Time Warp was able to speedup the simulation by a factor of 28.6 on 60 Mark III processors. STB88 ran out of memory with three (and two) processors, although the sequential simulation run completed on one processor. Mark III speedup curve exhibits an S-shaped curve above 48 processors, this degradation is attributed to poor load balancing. A different version of STB88 produced a speedup of 36.8 on 100 Butterfly processors.

Fujimoto conducted several queuing network simulations[2,3] using direct cancellation. He reported speedups as high as 57, using 64 processor BBN Butterfly. Figure 6.9(a)[2] shows the performance of Time Warp for First Come First Serve (FCFS) queues and queues with preemption (for different message population) as the number of processors is varied. The simulated queuing networks are closed and configured in a hypercube topology. A fixed number of jobs randomly circulate throughout the network. The routing at each hypercube node is random using a uniform distribution. Fujimoto also attempted to compare Time Warp to conservative simulation for the same types of queuing

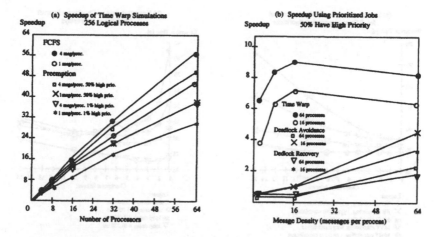

Figure 6.9: (a) [2]Speedup of 256 process hypercube simulator.
(b) [2]Speedup of the queuing network simulator using first-come first-serve queues.

network previously described. Figure 6.9(b)[2] shows that for low to moderate message density, Time Warp simulator far outperforms the conservative approaches. However, when message density is high, we can see that the performance of conservative algorithms begins to approach that of Time Warp. This is because when message population is high, it is very likely that all incoming links to one process will have at least one message, in this case the conservative algorithm operates at peak efficiency. For large hypercubes, Time Warp can extract more parallelism and hence performs better. However, for larger hypercubes conservative approaches perform worse since the message population per process decreases. Another important remark is that Time Warp performance degrades for large message populations because the time required for event list insertions dominates (the conservative algorithms use FIFO queues so they do not suffer from this problem).

Little work has been done to examine the trade-off between execution time and memory space in optimistic simulation. Preiss, MacIntyre and Loucks[23] reported some empirical results on simulation of closed stochastic queuing networks. They showed that the time-optimal and space-optimal checkpoint intervals are not the same. The checkpoint

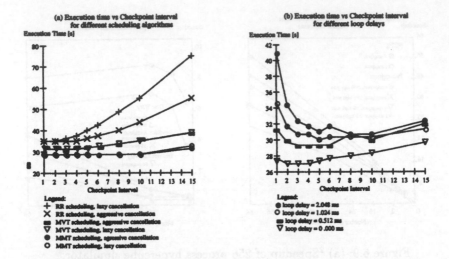

Figure 6.10: (a) [23]Execution time vs. Checkpoint interval, Procs=8
(8 LPs/Proc.), hypercube topology, load = 4, loop delay = 0 s. (b)
[23]Execution time vs. Checkpoint interval, processors=8
(8 LPs/Proc.), hypercube topology, load = 4, MMT scheduling,
lazy cancellation.

interval is defined as the number of input messages processed between
two checkpoint operations.

A checkpoint interval that is too small affects space more than time;
while a too large checkpoint interval affects time more than space. Figure 6.10(a)[23] shows the execution time versus the checkpoint interval
for an eight processor hypercube topology with a load (time-averaged
number of customers in each node) of 4, and zero loop delay (loop delay
is introduced to vary the checkpoint time at will). Different scheduling
policies (as discussed earlier) are considered. Figure 6.10(b)[23] shows the
execution time versus the checkpoint interval for different loop delays.
An important observation is that as the checkpoint time is increased,
the time-optimal checkpoint interval also increases. In other words,
when the checkpointing is costly one should save state less frequently.
Finally, Figure 6.11[23] shows that increasing the checkpoint interval from
one to four substantially decreases the maximum state storage. However, increasing the checkpoint interval beyond four has little added

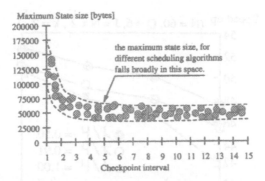

Figure 6.11: [23]Maximum state size vs. Checkpoint interval, Procss=8, hypercube topology, load = 4, loop delay = 0s.

benefit.

Madisetti, Hardaker and Fujimoto[18] implemented the MIMDIX Operating System 1.0, based on the Probabilistic Synchronization (PS), on the 32-processor BBN Butterfly system. They also reported their first performance measurements of PS on the simulation of a open network of communicating processes. Initial results show a significant improvement in speedup over Time Warp (TW). The application of concern is a open network of processes, each of which is able to communicate with one or more processes, update its local clock, and move forward in simulated time. The first generation of TW[14] was not designed to simulate open networks and could not carry out the parallel simulation (Memory exhaustion). Hence flow control based on blocking was implemented in Time Warp; this version is referred to as TW-1. In MIMDIX, the probability of the SYNC (p_s), and the delay δ were both varied to determine the optimal values of speedup and efficiency. Results show that for balanced systems, efficiency as low as 10 % observed in TW-1 could be boosted up to 60 % with MIMDIX (speedups ranging from 12 to 18 for a total number of processors equal 24). For unbalanced systems, the performance of MIMDIX is about twenty times better than TW-1. In other set of experiments, TW-1 was optimized by blocking the source from pumping messages until and unless the timestamps of messages were close to the GVT of the system. TW-1 with this optimization reached an efficiency of about 80-95 % of that of MIMDIX. However, TW-1 requires control of messages not only from outside the

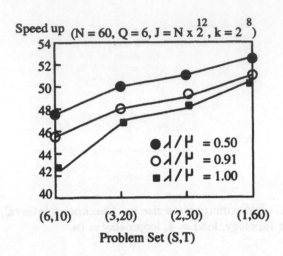

Figure 6.12: [15]Feed Forward Networks: Space-Time Speedup.

network, but also within the system. It is expected that Time Warp with cancelback will meet these requirement.

Bagrodia, Chandy and Liao[15] reported interesting performance results in the simulation of Feed Forward Networks (FFN) using the space-time algorithm (Chandy-Sherman[7]). Figure 6.12[15] shows the speedup as function of the space-time partition of a FFN with 6 stages (S), a fanout of 2, and a server queuing space (Q) of 6. N is the number of Processors used in the parallel simulation, J is the total number of jobs that were serviced, λ is the job arrival rate, μ is the service rate at each server, and k represents the number of jobs that are processed at a stage before they are forwarded as a packet to the next stage. A space-time mapping (S, T) means that the degree of state decomposition is S, while the degree of time decomposition is T. As shown by Figure 6.12, the overall speedup is best when the entire network is executed using only time parallelism.

6.5. Conclusion

Much is known about the time performance of parallel simulation. However, little work has been done to analyze the trade-off between time and space. Cancelback and artificial rollback, as memory management of the future for Time Warp, have been shown to be optimal when implemented in shared memory environment. No time performance of these cancel back algorithms has been reported, except when implemented as artificial rollback. Time Warp is efficient when the degree of parallelism in the system offsets the overhead in rollback and recovery. Memory overhead continues to be the most important source of performance degradation. Different checkpointing frequencies have been attempted; results show that time-optimal and space-optimal frequencies are not necessarily the same. It is expected that hardware support will keep state saving overhead to a manageable level. Analytic models, depend critically on the assumptions made on the synchronization cost (i.e., rollback cost), and can often predict contradicting results. Nevertheless, analytic models can still provide some useful insights to the efficiency of the implementation. Lookahead appears to be essential to obtain significant speedups, especially in conservative methods. First-come-first-serve queues are suited for conservative simulation algorithms because they contain a high degree of lookahead. However, the introduction of characteristics such as preemptive behavior and prioritized jobs, can reduce the lookahead considerably and cause a degradation in performance of conservative simulation. Probabilistic Synchronization provides the system designer with a number of control features, which when chosen appropriately can result in good performance and acceptable memory utilization. This has been demonstrated via the MIMDIX Operating System that is currently operational on the BBN Butterfly parallel computer. Space-time simulation is still another promising approach which invites the implementor to make use of both space and time parallelism along with a converging iterative algorithm.

An important factor affecting the performance of parallel simulation is the assignment of processes to processors; a poor partitioning can be very inefficient. On the other hand, if there exists a sufficiently large number of processes relative to the number of processors, efficiency is nearly always very high.

REFERENCES

[1] Fujimoto, R.M., "Parallel Discrete Event Simulation." *Communication of the ACM* , pp. 31-53 , October 1990.

[2] Fujimoto, R.M., "Time Warp on a Shared Memory Multiprocessor." *Trans. Soc. for Compu. Simul.*, pp. 211-239, July 1989.

[3] Fujimoto, R.M., "Performance of Time Warp under Synthetic Work-loads." *Proceedings of the SCS Multiconference on Distributed Simulation 22, 1*, pp. 23-28, January 1990.

[4] Fujimoto, R.M., Tsai, J., and Gopalakrishnan, G., "Design and Performance of Special Purpose Hardware for Time Warp." *Proceedings of the 15th Annual Symposium on Computer Architecture*, pp. 401-408, June 1988.

[5] Chandy, K.M., and Misra, J."Distributed Simulation: A Case Study in Design and Verification of Distributed Programs." *IEEE Trans. on softw. Eng.*, pp. 440-452 , September 1979.

[6] Chandy, K.M., and Misra, J."Asynchronous Distributed Simulation via a Sequence of Parallel Computations." *Communication of the ACM*, pp. 198-205 , November 1981.

[7] Chandy, K.M., and Sherman, R. "Space, Time and Simulation " *Proceedings of the SCS Multiconference on Distributed Simulation*, pp. 53-57, March 1989.

[8] Gafni, A. "Rollback Mechanisms for Optimistic Distributed Simulation Systems." *Proceedings of the SCS Multiconference on Distributed Simulation*, pp. 61-67, July 1988.

[9] Misra, J. "Distributed Discrete Event Simulation." *ACM comput. surv.*, pp. 39-65, March 1986.

[10] Bryan, R.E. "Simulation of Packet Communications Architecture Computer Systems." *MIT-LCS-TR-188, Massachusetts Institute of Technology*, 1977.

[11] Jefferson, D.R. "Virtual time." *ACM Trans. Prog. Lang. and Syst.*, pp. 404-425, July 1985.

[12] Jefferson, D.R., and Sowizral, H. "Fast Concurrent Simulation using the Time Warp Mechanism, part I: Local Control." *Tech. Rep. N-1906-AF, RAND corporation*, December 1982.

[13] Jefferson, D.R. "Virtual time II: the Cancelback Protocol for Storage Management in Time Warp." *Proceedings of the 9th ACM Symposium on Principles of Distributed Computing*, pp. 75-90, August 1990.

[14] Jefferson, D.R. et al, "The Time Warp Operating System." *Proc. of the 11th Symp. on Operating Systems Principles*, pp. 77-93, November 1987.

[15] Bagrodia, R., Chandy, K.M., Liao, W.-T., "An Experimental Study on the Performance of the Space-Time Simulation Algorithm." *Proceeding of the SCS Multiconference on distributed simulation*, pp. 159-168, January 1992.

[16] Madisetti, V., "Self-Synchronizing Concurrent Computing Systems." *UCB ERL/M89/122, Univ. of California at Berkley*, 1989.

[17] Madisetti, V., Hardaker, D., "Synchronization of Distributed Data-Driven Computation." *ACM Trans. on Model. and Comp. Sim.* Vol 2, No 1, pp 11-51, January 1992.

[18] Madisetti, V., Hardaker, D., "The MIMDIX Operating System for Parallel Simulation." *Proceeding of the SCS Multiconference on Distributed Simulation*, pp. 65-74, January 1992.

[19] Madisetti, V., Warland, J., and Messerschmitt, D., "Wolf: a Rollback Algorithm for Optimistic Distributed Simulation Systems." *Proceedings of 1988 Winter Simulation Conference*, pp. 296-305, December 1988.

[20] Madisetti, V., Warland, J., and Messerschmitt, D., " Asynchronous Algorithms for the Parallel Simulation of Event-Driven Dynamical Systems ", *ACM Transactions on Computer Modeling and Simulation*, Vol 1, No 3, pp. 244-274, July 1991.

[21] Felderman, R.E, and Kleinrock L., "Two Processor Time Warp Analysis: Some Results on a Unifying Approach." *Proceeding of the SCS Multiconference on Distributed Simulation*, pp. 3-10, January 1991.

[22] Felderman, R.E, and Kleinrock L., "Two Processor Conservative Simulation Analysis." *Proceeding of the SCS Multiconference on Distributed Simulation*, pp. 169-177, January 1992.

[23] Preiss, B.R., MacIntyre, I.D., Loucks, W.M., "On the Trade-off between Time and Space in Optimistic Parallel Discrete-Event Simulation." *Proceeding of the SCS Multiconference on Distributed Simulation*, pp. 33-42, January 1992.

[24] Lin, Y.-B., and Preiss, B.R., "Optimal Memory Management for Time Warp Parallel Simulation." *ACM Transactions on Computer Modeling and Simulation*, Vol 1, No 4, pp. 283-307, October 1991.

[25] Lin, Y.-B., "Memory Management Algorithms for Optimistic Parallel Simulation ." *Proceeding of the SCS Multiconference on Distributed Simulation*, pp. 43-52, January 1992.

[26] Lin, Y.-B., and Lazowska, E., "Reducing the State Saving Overhead for Time Warp Parallel Simulation." *Tech. Rep. 90-02-03, Dept. of Computer Science, University of Washington, Seattle, Washington*, February 1990.

[27] Chandak, A., "Vectorization of Discrete Event Simulation." *Proceedings of the 1983 International Conference on Parallel Processing*, pp. 359-361, August 1983.

[28] Berry, O. "Performance Evaluation of the Time Warp Distributed Simulation Mechanism." *Ph.D. Thesis, University of Southern California*, May 1986.

[29] Sokol, L.M., Briscoe, D.P., and Wieland, A.P. "MTW: a strategy for scheduling Discrete Simulation Events for Concurrent Execution." *Proceedings of the SCS Multiconference on Distributed Simulation*, pp. 34-42, July 1988.

[30] Som, T.K., Cota, B.A., and Sargent, R.G. "On Analysing Events to Estimate the Possible Speedup of Parallel Discrete Event Simulation." *Proceedings of the 1989 Winter Simulation Conference*, pp. 729-737, December 1989.

[31] Lavenberg, S., and Muntz, R. "Performance Analysis of a Rollback Method for Distributed Simulation." *Performance '83. Elsevier Science Pub., North Holland*, pp. 117-132, 1983.

[32] Mitra, D., and Mitriani, I. "Analysis and Optimum Performance of Two Message-Passing Parallel Processors Synchronized by Rollback." *Performance '84. Elsevier Science Pub., North Holland*, pp. 35-50, 1984.

[33] Lipton, R.J., and Mizell, D.W. "Time Warp vs. Chandy-Misra: a Wost-Case Comparison." *Proceeding of the SCS Multiconference on Distributed Simulation*, pp. 137-143, January 1990.

[34] Nicol, D.M. "Performance Bounds on Parallel Self-Initiating Discrete-Event Simulations." *Tech. Rep. 90-21, ICASE*, March 1990.

[35] Presley, M., Ebling,M., Wieland, F., and Jefferson, D.R. " Benchmarking the Time Warp Operating System with a Computer Network Simulation." *Proceeding of the SCS Multiconference on Distributed Simulation*, pp. 8-13, March 1989.

[36] Wieland, F., Hawley, L., Feinberg, A., Dilorento, M., Blume, L., Reiher, P., Beckman, B., Hontalas, P., Bellenot, S., and Jefferson, D.R., "Distributed Combat Simulation and Time Warp: The Model and its Performance." *Proceedings of the SCS Multiconference on Distributed Simulation*, pp. 14-20, March 1989.

[37] Turner, S.J., and Xu, M.Q., "Performance Evaluation of the Bounded Time Warp Algorithm." *Proceedings of the SCS Multiconference on Distributed Simulation*, pp. 117-126, January 1992.

[38] Wagner, D.B., Lazowska, E.D., and Bershad, B.N., " Techniques for Efficient Shared-Memory Parallel Simulation." *Proceedings of the SCS Multiconference on Distributed Simulation*, pp. 29-37, 1989.

[39] Walrand, J.C., "An Introduction to Queueing Networks", *Prentice-Hall, NJ*, 1988.

CHAPTER 7

LOAD BALANCING STRATEGIES FOR PARALLEL SIMULATIONS ON A MULTIPROCESSOR MACHINE†

Azzedine Boukerche and Sajal K. Das

7.1. INTRODUCTION

Research on parallel simulation has been stimulated during the last decade by excessive computational needs of large scale applications. Such applications concern mainly the verification of the design and the performance of complex systems such as computer and communication networks, and VLSI circuits, to mention a few. Accordingly, several methods for parallel simulation have been proposed. These techniques for parallel simulation can be classified into two groups, *conservative* approach and *optimistic* approach. While conservative synchronization techniques rely on *blocking* to avoid violation of dependence constraints, *optimistic* methods rely on detecting synchronization errors at run-time and on recovery using a *rollback* mechanism. Despite the fact that research on parallel simulation has been ongoing for the past decade, the problem of partitioning/mapping and load balancing for parallel simulation has been the subject of limited attention.

In this chapter, we consider the problem of load balancing for parallel simulation for execution on multiprocessor machines. We describe

---

⁰†This work was in part supported by Texas Advanced Technology Program grant TATP-003594031, and a grant from Nortel at Richardson, Texas. Email Address: {azedine,das}@ponder.csci.unt.edu

several load balancing techniques for both conservative and optimistic synchronization protocols.

Load balancing is an important component in improving the efficiency of parallel simulations, because it distributes an even workload over all processors. As we shall see in this chapter, an even workload over all processors results in significantly reducing the excution time of simulations running under two (diametrically opposite) classes of algorithms for the synchronization of parallel simulation –the conservative approach inspired by Chandy, Misra and Bryant[34], and the optimistic approach pioneered by Jefferson[23]. Hence, the load balancing problem is approached in two different ways. First, we consider the conservative synchronization protocol that makes use of Chandy-Misra null messages[10]. Next, we consider the optimistic approach for parallel discrete event simulation in which logical processes (LPs) proceed with their computations without any constraints.

The remainder of this chapter is organized as follows. In the next section, we briefly review previous and related work. Section 7.3 describes several partitioning strategies for conservative parallel simulations making use of Chandy-Misra null-message protocol. We propose an algorithm, that makes use of a simulated annealing technique with an adaptive search schedule to find good (sub-optimal) partitions. We discuss the algorithm, its implementation and report on the performance results of simulations of a partitioned FCFS queueing network model executed on a multiprocessor machine.

In Section 7.4, we consider the optimistic approach for parallel discrete event simulation in which the computing requirements of different LPs are not balanced, or if the processors are not homogeneous, some LPs may lag behind in simulation time while others surge forward. If the simulation clocks of different LPs are not progressing at the same rate, *cascading* rollbacks may occur nullifying the potential benefit of optimistic parallel simulation. Hence, it is necessary to balance the computational load on different LPs in such a way that their local simulation clocks advance at almost the same rate. We propose a dynamic load balancing algorithm in which load migration takes place between physical processors. It is based on the global virtual time (GVT) computation. The performance of this algorithm is analyzed probabilistically on a realistic model, which treats the progress of simulation clock (with the execution of events) as a random walk. Conditions are derived under which the proposed solution is effective, and bounds on the rate of progress of simulation are computed.

Finally, in Section 7.5, a summary of the chapter and a brief account of future extensions to our work are presented.

7.2. PREVIOUS AND RELATED WORK

The partitioning and the load balancing problems in parallel and distributed systems have received quite a lot of attention in the past decade[4,36,46]. Approaches to partitioning have also been applied to various VLSI design problems [31]. However, in these many formulations of partitioning and load balancing problems, the problem of finding an optimal partitioning is found to be NP-hard[18] in all but very restricted cases. Thus research has focused on the development of heuristic algorithms to find suboptimal partitioning solutions. Much of the research on partitioning and mapping problems can be classified as either graph theoretic[4], queueing theoretic[26], based on mathematical programming[20], numeric and non-numeric heuristics[40], and combined methods.

In terms of the partitioning problem and system graph, the problem can be stated as grouping the nodes of the system graph into clusters so that the sum of link weights external to the clusters is minimized, and the execution load is distributed uniformly among the processors. We assume that the inter-communication cost (IMC) between each pair of processes is known, represented by the weight of an undirected edge connecting the two nodes (An IMC cost of zero means no communication takes place between the two processes, and therefore they are not connected in the graph.) and the processor utilization is maximized. Bokhari[4] and Nicol[40] discuss optimal solutions for restricted cases of the problem. They provide polynomial partition and mapping a chain of modules to heterogeneous systems, and thus their scope is limited to programs that can be represented as a chain of modules.

Kim and Brown[28] consider the problem of task allocation in distributed systems for a special case of clustering called *linear clustering*, i.e., no cluster contains any pair of independent tasks. Sarkar[46] presented another heuristic based on an unbounded number of processors. The resulting clustering is not necessarily linear, therefore the algorithms belong to the class of nonlinear clustering algorithms. Wu and Gaski[53] developed another clustering heuristic and incorporated it into a programming tool for hypercube architectures. Each of these heuristics can be considered as a scheduling heuristic that employs the concept of clustering. In other words, if the obtained clustering is non-linear, some criterion is used to determine the order of executing the tasks of each cluster. Note that these heuristics assume a fully connected target machine and that the number of processors is always greater than or equal to the number of clusters.

Kling and Banerjee[27] have proposed simulated evolution as an alternative to annealing, and applied it in the context of cell placement in VLSI design. Their technique is the mathematical analog of the natural selection of biological environments, and it performs three basic steps: evaluation, selection, and allocation. The first step computes the "goodness" of the particular cell position. In the second step, the cells are probabilistically selected for replacement according to their goodness. Finally, the third step removes cells from their current allocation and searches for an improved location. These three steps are repeated until no further improvement is needed. The experimental results showed that the simulated evolution is slow. As a consequence the authors make use of hierarchical and window techniques that reduce the running time significantly.

Kernighan and Lin[25] propose a two-way partitioning algorithm with constraints on the final subset sizes. They applied pairwise swapping and iterated on all pairs of nodes to find the best improvement on the existing partition. Fiduccia and Mattheyses[15] further improved this algorithm by developing a clever implementation for each iteration to achieve a linear complexity for each iteration. Krishnamurthy[31] introduced the multiple level gain model for multi-pin nets in VLSI. Sanchis[44] then adapted this model to multiple-way partitioning. Her algorithm attempts to minimize the communication between processors and keep the number of processes per processor within a specified range. The algorithm uses a concept of levels; each level successively produces a better cut set. A VLSI component model and a network of SUN-workstations were employed for experimentation.

The Kernighan-Lin based algorithms unfortunately share the common weakness that they are often trapped by local minima when the size of the problem is very large. One way to overcome this difficulty is to form clusters, and then condense these clusters into single nodes prior to the execution of the Kernighan-Lin based algorithms. The complexity of the problem is thus dramatically reduced, which in turn improves the performance of the partitioning algorithm [9].

However, each of these preceding approaches to the partitioning problem provides good solution for restricted applications, and they are not suitable for parallel simulation since the synchronization constraints exacerbate the dependencies between the LPs. In order to achieve the best performance, the partitioning and the load balancing should decrease the running time of the simulation compared to a random partitioning. However, it is not yet clear how to measure the parallelism, or how to extract it from a simulated system in advance.

7.2.1. Parallel Time-Stepped Simulation

Nicol and Reynolds[37] propose a "statistical approach to dynamic partitioning" for a homogeneous system of K processors. They present a statistically based method for both the static and the dynamic partitioning of logic networks for parallel simulation. They investigate a *time-step* simulation of VLSI circuits with no feedback. Circuit (gate) delays are determined by statistical analysis. A clustering algorithm is applied to the process graph to distribute the processes among the K processors. The decision to re-compute (adapt) the partition is made by an "identification technique" which employs histories to determine if the cycle times have changed. If so, a new partition is computed. A Bayesian decision process is used to decide whether the new partition should be adopted or not. The number of cycles remaining in the simulation along with the cost of adopting the new partition is considered in the Bayesian decision. Their algorithms have been tested empirically, and each has been found effective. However, their algorithms are limited to a network that has no feedback.

Nicol[39] considers the decision problem regarding the remapping of workload to processors in a parallel time-stepped simulation where the behavior of the computation is characterized as a sequence of phases, the workload is uncertain, phases exhibit stable execution requirements during a given phase, but requirements may change radically between phases. The key issue for the remapping decision problem is seen to be the relatively accurate assessment of when remapping will lead to performance gain. The decision to re-compute (adapt) the partition is made by a threshold decision heuristic based on simple calculation of apriori probability (p_n) that a phase change occurs by a step n. As a benchmark, he investigates the battlefield simulation. His results indicate that the Bayesian updating of p_n combined with a reasonably high decision threshold provides good protection against premature remapping.

7.3. CONSERVATIVE PARALLEL SIMULATION

Parallel discrete event simulation models physical systems as a collection of logical processes (LP) that communicate only via time stamped event messages. In a conservative paradigm, an event cannot be simulated by an LP before it is certain that an event with a smaller time stamp cannot arrive. As a consequence of this blocking behavior, deadlocks arise. Several solutions to this problem exist, each requiring a certain amount of overhead[7,17,34]. A substantial amount of work has been done to evaluate the performance of these strategies[16,32,38]. However, the problem of load balancing for conservative parallel simulation has been the subject of limited attention. Load balancing methods are typically classified into two major groups: *static* and *dynamic* load balancing algorithms.

In a static load balancing (often referred to as partitioning/mapping), the assignment of processes (LPs) is done before program execution begins. The goal of static partitioning is to minimize the overall execution time of the simulation while minimizing the communication overhead. The majors advantage of static partitionning/mapping is that all overhead of the scheduling process is incurred at compile time, resulting in a more efficient execution time environement compared to dynamic load balancing. With this goal in mind, very few methods for balancing the load of parallel simulations have been proposed. Our objective in this section is to study the potential importance of partitioning on the performance of conservative parallel simulations. We focus upon Chandy-Misra null-message synchronization mechanism as the underlined model, described next.

7.3.1. Chandy-Misra Null Message Protocol

Chandy and Misra[10] employ null messages in order to avoid deadlocks and to increase the parallelism of the simulation. When an event is sent on an output link a null-message bearing the same time stamp as the event message is sent on all other output links. As is well known, it is possible to generate an inordinate number of null messages under this scheme, nullifying any performance gain[16].

In order to increase the efficiency of this basic scheme, we employ the following approach[7] . In the event that a null-message is queued at an LP and a subsequent message (either null or event) arrives on the same channel we overwrite the (old) null message with the new message. We associate one buffer with each input channel at an LP to store null messages, thereby saving space as well as the time required to perform the queueing and de-queueing operations associated with null messages.

7.3.2. Iterative Improvement Based Partitioning Algorithm

Nandy and Loucks[35] presented a static partitioning algorithm for conservative parallel logic simulation. The synchronization protocol made use of null messages[10]. A min-cut iterative improvement algorithm, similar to Fiduccia and Mattheyes[15], has been used to minimize the communication overhead and to uniformly distribute the execution load among the processors. The authors[35] assumed that there are as many clusters as there are processors, each cluster is assigned to one processor.

The algorithm starts with an initial random partition, and then iteratively moves processes between clusters until no improvements can be found (a local optimum). The processes are moved among the clusters, so that the total cut-size is reduced and the cluster sizes remain balanced. All possible moves for each process are considered, and the process which contributes to the maximum gain is chosen. A process is moved only if it does not violate the block size constraints. As a benchmark, they use the simulation of circuits modeled at the gate level. A message passing multicomputer composed of eight T-800 INMOS transputers was employed as the simulation platform. They report a 10-25% reduction in simulation time from the simulation time of a random partition.

7.3.3. Adaptive Partitioning Based on Simulated Annealing

A new approach to the approximate solution of hard combinatorial optimization problems such as graph partitioning has been proposed by Kirkpatrick, et al[29]. This approach, known as simulated annealing (SA), is based on ideas from statistical mechanics and motivated by an analogy to the behavior of physical systems in the presence of a heat bath. Simulated annealing is a probabilistically directed iterative improvement scheme. While iterative improvement accepts a new configuration of lower cost and rejects more costly states, SA escapes from local minima by sometimes accepting higher cost arrangements with a probability determined by the simulated 'temperature'. It has been known that simulated annealing is a valuable contribution to the field of heuristic methods, and this approach has several advantages over other for solving combinatorial optimization problems. First, it is problem-independent. By substituting a few problem specific data structures and functions, the simulated annealing algorithm can be applied to many combinatorial optimization problems[24]. Second, simulated annealing can easily handle multiple, potentially conflicting goals of a problem. It has been successful in solving combinatorial problems such as cell placements[30], floorplan design[52] and task assignments[51].

Boukerche and Tropper[5] recently discussed the application of simulated annealing with an adaptive schedule to the partitioning problem. Their objective is to develop a partitioning algorithm and is based upon realistic estimates of computation and communication load. We present this scheme in this section.

The structure of the algorithm is contained in Fig. 7.1. In the application of partitioning, the state is represented as the assignment of each process (LP) to a partition, which in turn is assigned to a processor.

Starting from an initial partition, a *move* generates another partition. An objective function (F) evaluates the quality of this partition. This new partition is accepted if it is better, or probabilistically accepted if it is worse. The probability of accepting a new partition differing by $\Theta(F)$ is denoted as $\mathcal{P}(\Theta(F))$. In the experiments, a Boltzmann distribution is used. $\mathcal{P}(\Theta(F)) = min(1, e^{-\frac{\Theta(F)}{kT}})$. This function accepts all changes of $\Theta(F) \leq 0$ with $\mathcal{P}(\Theta(F))$. It is assumed that the Boltzmann constant k equals to 1 in order to avoid inconveniently large temperature scales.

When an equilibrium state, represented as the assignment of each process (LP) to a partition has been reached, the temperature is decreased. This iteration is repeated until the algorithm meets a termination condition or until the temperature reaches its lowest value T_{final}. The termination condition and the equilibrium condition together are

Initialization :
Choose α (the annealing factor), $T_{initial}, T_{final}$.   (* $T_{initial}$ and T_{final} are the
initial and final temperature values *)
   Choose an initial partition $S_{current}(X) = (x_1, ..., x_n)$, where x_i represents the
partition
to which process LP_i is assigned.
Evaluate $Cost_{current}(X) = F(X)$;   (* We describe the choice of F below *)
$T \leftarrow T_{initial}$;
While $(T > T_{final})$ and (Termination Condition is not met)
  While (Equilibrium Condition is not met)   (* We describe the termination and
                                        equilibrium conditions below*)
    $S_{new}(X') \leftarrow$ **Move**$(S_{current}(X)$;   (* We describe the move strategy
                                              below *)
    $Cost_{new} \leftarrow EvaluateF(S_{new}(X'))$;
    Compute $\Theta(Cost) = Cost_{new} - Cost_{current}$;
    If $(\Theta(Cost) \leq 0)$ or $(e^{\frac{-\Theta(cost)}{T}} <$ **random**$())$
      $S_{current}(X) \leftarrow S_{new}(X')$;   (* Accept the new solution *)
      $Cost_{current} \leftarrow Cost_{new}$;
    EndIf
  EndWhile
  $T \leftarrow$ **Next_Temp(T)**;   (*We describe the cooling schedule below *)
EndWhile

Fig. 7.1. A Simulated Annealing Algorithm

called the *annealing schedule.*

Boukerche and Tropper[5] presented an objective function (F) to evaluate the quality of the partitioning solution generated by the algorithm.
The function was chosen such that the inter-processor communication
conflicts are minimized, processor load remains balanced, and the probability of sending a null-message between processors is minimized.

We define expressions for these parameters as well as the objective
function F which represents the quality of the partitioning

$$F = \mathcal{F}(LOAD, Diameter_{Avg}, IPC, IPL).$$

Let us denote by λ_{ij} the average traffic between each pair of adjacent
processes LP_i and LP_j, n_i the number of processes assigned to processor $Proc_i$, N_{links} the total number of links, N_{tot} the total number of
processes (or nodes) and $\mathcal{K}$ the number of processors.

We define the load at i-th processor $(Proc_i)$ as follows:

$$Load_i = \sum_{j} \sum_{p_k \in Proc_i} \lambda_{jk} * (serv_k + T_{send} + T_{rcv})$$

where $serv_k$ is the amount of time LP_k takes to process an event, T_{send} is the amount of time it takes for an LP to send an event, and T_{rcv} the amount of time it takes for an LP to receive the event. If the event is generated by an LP in processor $Proc_i$, then T_{send} and T_{recv} are negligible.

We wish for $Load_i$ to be approximately equal to $\frac{1}{\mathcal{K}}\sum_{j=1}^{\mathcal{K}} Load_j$ and hence define

$$LOAD = \mathcal{K}^{\mathcal{K}} \prod_{i=1}^{\mathcal{K}}(Load_i/Load_{total})$$

where $Load_{total} = \sum_{i=1}^{\mathcal{K}} Load_i$. Here $\mathcal{K}^{\mathcal{K}}$ is a normalization factor that causes the maximum of $LOAD$ to be 1. If the number of LPs within each processor is equal, i.e., $Load_i = \frac{1}{\mathcal{K}}.Load_{total}$, the quantity $LOAD$ reaches a maximum equals to 1. Therefore, the goal is to maximize the quantity $LOAD$.

We also consider the quantity: $Diameter_{avg} = \frac{1}{\mathcal{K}}.\sum_{i=1}^{\mathcal{K}} Diameter_i$ as an average measure of distance within each cluster where $Diameter_i = \max_{j,k \in Proc_i}(Dist_{jk})$, and $Dist_{jk}$ is the number of hops between LP_j and LP_k mapped to the processor $Proc_i$. LPs that are close to each other are expected to communicate more than if they are far away from each other (this is a characteristic of traffic on computer networks and telephone systems). Hence assigning them to the same processor is expected to reduce the running time of the simulation.

The relative inter-processor communication factor is defined as follows:

$$IPC = \frac{t_{comm}}{t_{calc}} . \sum_{i,j} \lambda_{ij}.Cost_{ij} \qquad (1)$$

where t_{comm} is the mean time to send one message between two processors, t_{calc} is the mean time for the processing of an event message, and $Cost_{ij}$ is the communication overhead between processes LP_i and LP_j. For example, the communication overhead for the hypercube architecture is equal to $T_{startup}(n) + nT_{send}(n) + (h-1)T_{hop}(n)$, where n is the number of bytes in the message, h is the number of hops the message must traverse, $T_{startup}(n)$ is the time required to prepare the message for transmission, T_{send} is the time required to transmit a message and $T_{hop}(n)$ is the communication delay incurred at each hop beyond the initial one.

The ratio t_{comm}/t_{calc} reflects the relative costs of inter-node communication and computation. Throughout our discussion, we assume that the communication cost between two processes assigned to the same processor is negligible. The relative inter-processor equation (1) has been chosen because it is more expensive to send a message between processors than to do so within a single processor. It is assumed that we have estimates for λ_{ij} for each pair of adjacent processes LP_i and LP_j.

In order to reduce the number of null messages (IPL), we minimize the number of links between each pair of processors, $Proc_i$ and $Proc_j$. Barnes[1] merge the two partitions with the largest number of interconnections into a new partition if an upper bound on the number of LPs in a processor is not exceeded. While this approach generates partitions with a small number of interconnections, the number of LPs per processor can be quite different. Sporrer and Bauer[47] suggested the following ratio: $k_{ij} = m_{ij}/(\sum_l m_{il})$ where m_{ij} represents the number of interconnection between $Proc_i$ and $Proc_j$ and m_{ii} is set to zero. Better partitions[47] were obtained making use of k_{ij} instead of m_{ij}. Bearing this in mind[5], we chose to define $IPL = \sum_{ij} k_{ij}$

A good partitioning is one which meets the criteria of maximizing the $LOAD$ and minimizing the quantities: IPC, $Diameter_{avg}$, and IPL. Boukerche and Tropper[5] expressed the basic problem as finding a partition which maximizes the following quantity

$$F = Log \left(\frac{LOAD^\alpha}{(1 + IPC)^\beta.(1 + IPL)^\gamma.(1 + Diameter_{Avg}^\delta)} \right)$$

where α, β, γ and δ are the relative weights of the corresponding parameters. The factors in the denominator are of the form $(1 + x)^y$ because of the possibility that x might approach zero. The Log function is used to avoid inconveniently large factor scales. In the experiments, $\alpha = \beta = \gamma = \delta = 1$ yielded good results (see experimental results).

The algorithm starts with an initial random partition and then iteratively moves processes between clusters until the algorithm reaches an *equilibrium state*.

At each temperature T, the algorithm is said to be at *equilibrium* if the number of acceptances exceeds a constant β, or if the number of rejections at temperature T exceeds $\gamma = f(\beta)$. The value of β is decreased at successive temperatures, thereby causing the system to attain equilibrium to be approximately equal for all temperatures. We consider the following expressions: $\beta_n = a.\beta_{n-1}$ and $\gamma_n = r.\beta_n$; where $0 \leq a < 1$ and $r \geq 100$. The reason for choosing $\gamma_n > \beta_n$ is that as the temperature is lowered, the number of rejections increases. If the partitions computed cannot be improved over ψ consecutive temperatures, the algorithm terminates. The choice of r, a and ψ were determined as a result of our experiments[5].

In order to decrease the number of iterations and (thus) the running time of the algorithm, we decided to use an adaptive schedule. The motivation to derive an adaptive schedule is based on the observation that the behavior of SA is very different at high and low temperatures. Indeed, at high temperatures, the number of acceptances dictates equilibrium, while at low temperatures the number of rejected states dictates equilibrium. Another important factor is the number of iterations required to reach the final state. Experimental results indicate that using an adaptive cooling schedule produces satisfactory results. The cooling strategy is based on the following: (1) Allow a few iterations in which virtually every new state is accepted and where T is reduced quite rapidly from iteration to iteration; (2) Having left the high T regime, reduce T in such a manner that $\Theta(F)$ is approximately the same from iteration to iteration; (3) When T is reduced below a certain fixed apriori temperature ($T_{min} = 1.5$) then reduce T very rapidly so as to converge to a sub-optimal solution. The cooling schedule is shown below.

α	0.95	0.9	0.85	0.80	0.10
T	150	100	50	10	1.5

α Vs T

7.3.3.1 Experimental Results

The goal of the experiments conducted by Boukerche and Tropper [5] was to study the impact of the adaptive partitioning algorithm on the performance of a conservative synchronized parallel simulation making use of null messages[10]. The experiments have been conducted on an Intel iPSC/860 hypercube, which is a distributed memory multiprocessor, in which the processing elements are connected in a hypercube topology.

In the experiments, a queueing network model, in the form of a torus with an FCFS service discipline, was selected as a benchmark. Earlier simulation studies[16,32,38] showed that the performance of a simulation strategy is sensitive to the topology of the simulated network. A toroid network topology was selected primarily because it provides a stress test for the algorithms as a consequence of its many cycles. Furthermore, it does not contain any inherent bottlenecks and it is widely used in the performance studies of parallel simulation models[7,16].

In this model, each network node is a server with infinite capacity input queues. A server, modeled by an LP, removes an event from one of the queues and starts service. In the experiments, the shifted exponential service time distribution has been selected (i.e., $0.1 - log(uni())$, where $uni()$ returns a random real number uniformly distributed between 0 and 1.). When service completes, the event is forwarded to one of its neighbors selected with equal probability. A null-message, bearing the same time stamp as the event message, is sent on all other output links. We selected an arbitrary number of messages (six on each link) with which to start the simulation in order to control the amount of available parallelism. The message population as well as the number of LPs determine the amount of parallel activity that can occur in the simulation.

We made use of 22 x 22 tori and varied the number of processors from 2 to 16. The experimental data were obtained by averaging several trial runs. We present the results below in the form of graphs of the execution time (in seconds) and the speedup as a function of the number of processors employed in the model. The speedup $SP(n)$ is calculated as the execution time ET_1 required for a sequential simulator to perform the simulation on one processor divided by the time ET_n required for the parallel simulation to perform the same simulation on n processors, i.e., $SP(n) = ET_1/ET_n$.

Let us now turn to the results. Figure 7.2 portrays the values for the execution time of the simulation model obtained using the random partitioning and the SA-partitioning algorithm. As we can see from the curves, the SA-partitioning algorithm exhibits a better performance over the randomly partitioned one. We observe an approximate 25% reduction in the execution time of the simulation using the SA-partitioning algorithm over the randomly partitioned one when 2 and 4 processors are used. Increasing the number of processors from 4 to 8 results in a reduction of the run time of the simulation model by

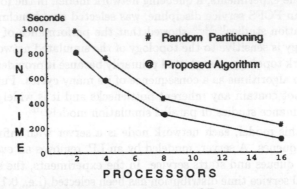

Fig. 7.2. Run Time Vs Number of Processors

approximately 25-35% using SA-partitioning strategy when compared to the randomly partitioned one.

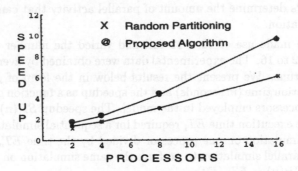

Fig. 7.3. Speedup Vs Number of Processors

Figure 7.3 portrays the speedup for both random, and SA-partitioning strategies. We observe significant speedups. The results show that careful static partitioning is important in the success of the Chandy-Misra null-message method[10]. The impact of partitioning increases with the number of processors. This is due to the fact that the communication cost increases with the number of processors.

7.4. OPTIMISTIC PARALLEL SIMULATION

In this section, we focus upon the time warp protocol for parallel discrete event simulation, in which each process continues to receive and execute the event messages out of its input queue until no messages remain or until a message arrives "in the past" (called a *straggler*). Upon receiving a *straggler*, the process execution is interrupted and a *rollback* action takes place. Rollback consists of restoring the process to the appropriate state and sending cancellation notices for messages produced by the rollback portion of the computation using *antimessages*. Excessive amount of rollbacks may cripple the whole system, without achieving any significant speedup.

To keep the number of rollbacks under control, Lubachevsky[33] introduced *filtered* rollbacks, which postpones execution of certain events optimistically. He also presented analytical evaluation using barrier synchronization scheme and defined two types of rollbacks—*cascading* and *echo*. Cascading may cause a very large number of rollbacks if the simulation system has a self feeding loop, a devastating effect. Several attempts to reduce the overhead cost of rollbacks have been reported[2,45].

An alternative approach is to use load balancing schemes to reduce the number of rollbacks. Little work has appeared to date on the problem of load balancing in optimistic approach[19,41,45].

7.4.1. Dynamic Load Management Based on Virtual Time

Reiher and Jefferson[41] presented prelimiary performance results of applying dynamic load management to Time Warp Operating System (TWOS), a special purpose virtual-based operating system that runs event driven simulations on parallel processors. Their approach is based upon the notion of *effective utilization*, which is the fraction of time the processor spends doing effective work. They used an estimator of effective work to perform dynamic load management. The estimator is produced by keeping track of how much processor time was used for each event. That amount is added into the phase's effective work estimator. Overhead doesn't count as effective work, nor does work that has been rolled back.

Their algorithm balances the effective utilization of the processors by using a bin-packing algorithm in conjunction with object migration. They apply their algorithm to a military simulation and a colliding puck simulation. Their preliminary results showed up to 25 % speedup when a random process allocation was employed. Although, the results presented do not show evidence of improvement of the run time of irregular simulations, yet, their results were quite encouraging.

7.4.2. Dynamic Load Management Based on Simulation Advance Rate

Glazer and Tropper[19] presented a dynamic load balancing algorithm, with an attempt to reduce the number of rollbacks through process migration and scheduling mechanisms. Their algorithm is based upon the notion of *simulation advance rate*. They define the simulation advance rate to be the rate at which a process advances its simulation clock as a function of the amount of CPU time it is given. This rate is dependent on the time required to execute an event, the amount of the time the simulation clock is advanced per event and the frequency of event arrivals. They attempt to minimize the difference between the local simulation clocks by controlling the amount of CPU time allocated to each process. Two versions of a load balancing algorithm refered to as *standard* and *dampened* algorithms, were presented which differ in their reaction to simulation changes. The former algorithm reacts quickly to any detected change in the simulation advance rates. The dampened algorithm takes a conservative approach to changes by employing a simple weighting mechanism. The load balancing algorithms were evaluated on the PARALLEX emulation tool. PARALLEX was used to emulate both shared and distributed memory environments with a variable number of processors. Several queuing network simulations were employed to evaluate the algorithms-a hierarchical, a pipeline, and a distributed network. Substantial speedups were produced by both versions of the algorithm. Simulation rates rose by a maximum of 71% and rollbacks were reduced by a maximum of 66%.

7.4.3. Dynamic Load Management Based on Processor Utilization

El-Khatib and Tropper[14] describe a distributed dynamic load balancing algorithm based on a token-passing mechanism. The algorithm makes use of the token employed in the GVT algorithm to gather data. The metric which they employ is processor utilization. Using the same networks as Glazer and Tropper[19], they obtain a 25% increase in the speed of the simulation compared to Time Warp. Experiments were carried out on a BBN butterfly. The algorithm, however, does not rely on the use of shared memory.

Avril and Tropper[22] describe a dynamic algorithm which attempts to distribute the load of a simulation evenly among the processors. Load is defined as the number of events processed since the last load balance. A dedicated processor is used to perform the load balancing. The algorithm was tested in the environment of logic simulation and was used in conjunction with Clustered Time Warp(CTW)[21]. CTW was developed principly for use with logic simulation. It allows rollbacks to occur only between clusters of processes which are transferred between processors. Two circuits having more than 20,000 gates were tested, and improvements of between 40% and 100% in the event throughput were noted.

7.4.4. Dynamic Load Management Based on The Rate of Progress of Simulation

Das and Sarkar[12] recently proposed a method to keep the local clocks of different LPs in step by balancing the load among different physical processors, and thereby reducing rollbacks in an optimistic simulation. This approach is based on the computation of the global virtual time (GVT). Using a random walk model, the authors derived conditions, under which the scheme is effective and computed bounds on the rate of progress of simulation. We present this dynamic scheme in this section. If a processor has more computing power, or if the logical processes (LPs) mapped onto a physical processor are not computation-intensive, some of the LPs from the heavily loaded physical processor are simply transfered. Before proceeding further, let us formally define GVT.

The *local virtual time*, lvt_i, of LP_i is defined to be the minimum among timestamps of all the events in the local event queue (E_i) of LP_i, i.e., $lvt_i = \min\{t(e_j) \mid e_j \in E_i\}$, where $t(e_j)$ is the time stamp of the event e_j. If the system adopts preemptive scheduling policy then $SC_i = lvt_i$, where SC_i is the local simulation time clock value of LP_i, denoting how far the simulation has processed at LP_i.

The *global virtual time*, $GVT(t)$, of the system at time t is defined to be the minimum among the virtual times at all nodes and the timestamp of all the transient messages. In other words, $GVT(t) =$ $\min\{\min\{lvt_i|$ for all $i\}$, $\min\{t(M_{i,j})|$ for all i,j such that LP_i communicates with LP_j $\}\}$. A message, M_{ij}, sent from LP_i to LP_j is said to be *transient* if it has been sent by LP_i but not yet received by LP_j. The time for transfering M_{ij} is given by $t(M_{ij})$.

Das and Sarkar[12] introduced a new parameter, called *boundary virtual time* (BVT), which is defined as the maximum of the simulation clock values of the LPs in the whole simulation system. The BVT denotes the minimum simulation time beyond which no LP has progressed. In other words, BVT is the same as the simulation clock value of the LP farthest ahead in simulation time. The computation of BVT may progress with GVT computation without additional overhead.

The difference, $Diff = BVT - GVT$ gives a measure of the asynchrony among the simulation clock values. The closer the value of $Diff$ to 0, the more synchronized are the different LPs in the system. The distribution function, $\mathcal{DF}(.)$ of the random variable $Diff$, is dependent on the application and the physical characteristics of the system, and it is somewhat difficult to characterize. However, if the application system and the simulation algorithm are known, an empirical estimation of this function is possible.

In this model, each LP_i maintains an outgoing message queue that stores the messages sent to other LPs. The messages in the outgoing queue may be differentiated from the ones already sent and the ones in the process of being sent. In the case of a rollback, the LPs only send antimessages corresponding to the messages already sent. An event queue contains the events or messages received from other LPs. LP_i goes through a number of states between the GVT and SC_i values, which are saved in a stack from which they can be retrieved and restored if needed. When a new GVT is computed, the states saved in the stack prior to GVT are discarded. Frequent state savings impose high overhead, which can be minimized by saving system states after a fixed number of state changes.

There is also a cost associated with every BVT and GVT computations. Hence, they cannot be computed too frequently. We assume that the GVT computation routine is invoked every T units of time. We also assume that there is a threshold parameter Δ such that if $BVT - GVT > \Delta$, only then the load balancing takes place. We will look into possible choices for T and Δ. Note that if the number of LPs mapped onto the physical processor holding the slowest LP is equal to one, the only way to speedup the simulation is to split the slowest LP into several smaller sub-lps. We assume that such a situation does not arise in our case.

Let us define the variables used in the following algorithm. Let LP_gvt (or LP_bvt) be the LP whose simulation clock value is the same as GVT (or BVT). A tie is broken arbitrarily. Let P_gvt (or P_bvt) denotes physical processor handling LP_gvt (or LP_bvt). The physical processor handling LP_i is denoted as $Map(i)$. The load balancing algorithm is formally described in Figure 7.4.

While stopping criteria is not met
    If (Current_Time $- T_{last} > T$)   (* T_{last}: Last time when GVT was comp
                          uted *)
        Compute GVT, BVT, LP_gvt and LP_bvt
        $S=\{LP_i$: so that either $(LP_i,$LP_gvt$)$ or $($LP_gvt$,LP_i)$ is an edge in
          LP-digraph$\}$;
        P_gvt $= Map($LP_gvt$)$;
        P_bvt $= Map($LP_bvt$)$;
        If (BVT$-$GVT $> \Delta$)
            Select LP_j other than LP_gvt from the set of LPs mapped onto P_gvt;
            migrate LP_j from P_gvt to P_bvt;
            (* Each processor should know about the LP assignment.*)
            Inform all LPs in S about LP_j migration by sending $Map(j)$;
        EndIf
        $T_{last} =$ Current_Time
    Else sleep
    Endif
EndWhile

Fig. 7.4. Procedure Load-Balance

We assume that the algorithm is executed by the operating system in the background. Now, instead of transferring one process, one can make it more general by transferring an arbitrary number of processes. But, then such a scheme may not guarantee substantial achievement unless the average number of LPs being simulated per physical processor is very large.

Each LP_i stores the value of the last GVT and the address (initially, this address is set to *nil.*) of the physical processor (if any) from where LP_i migrated after the GVT computation. This information is discarded after another GVT computation, if the simulation clock value at the time of migration was less than the new GVT. In the case of a rollback, we first search for the rolled back state in the current physical processor, $Proc_{Map(i)}$. If this state is unavailable, the address of the processor where the LP was assigned to, at the time of the rolled back state, is computed. The state vector of LP_i and messages sent after the rolled back time are also fetched. Therefore, we only need to transfer the unprocessed messages in the incoming queue, when an LP migrates. This way the cost of process migration is kept at its minimum. Though the entire approach is developed keeping the distributed memory models in view, the process migration can also be implemented on shared memory machines.

7.4.4.1 Performance Analysis of the Proposed Scheme

This section summarizes the results due to Das and Sarkar[12], which derives conditions under which the proposed scheme is effective, and we compute bounds for the rate of progress of simulation. Let Ψ be the total cost of a process migration. The following assumptions are also made about our system:

- Let $\mathcal{R}$ be the average cost of one rollback.
- Processor $Proc_i$ handles n_i LPs and $\sum_{i=1}^{K} n_i = n$, where K and n are respectively the number of physical processors and logical processes. Without loss of generality, we assume that LP_1 is the process being migrated from processor $Proc_1$ to $Proc_K$.
- The procedure starts with the zero value of the GVT. The simulation time is assumed to be discrete so that the value of the simulation clock always changes by an integer value.
- After the execution of an event, either the simulation clock jumps Z steps forward where Z follows a geometric distribution, or it jumps Z steps backward where Z follows a truncated geometric distribution[42]. Backward jump occurs only in a case of a rollback. The length of the backward jump at LP_i can be at most $SC_i - GVT$, where SC_i is the value of the simulation clock at LP_i. But for simplicity, let it range between 0 and SC_i. This assumption is justified by the fact that the tail of a geometric distribution decreases rapidly. If the difference between simulation clock and the last computed GVT is large enough or if the GVT is close to zero, this assumption will have negligible effect on the analysis. We can also attach a positive probability to the event that the amount of jump in simulation time is zero. This does not alter our analysis. If a is the probability of a rollback and p_1 and p_2 are the

parameters corresponding to the geometric distributions for forward and backward jumps, then the probability of forward jump is given by $\mathcal{P}(Z = k) = (1 - a)p_1 q_1^{k-1}$, for $k = 1, 2, 3, \ldots$, where $q_1 = (1 - p_1)$. The probability of backward jump is, $\mathcal{P}(Z = -k) = acp_2 q_2^{k-1}$, for $k = 1, 2, 3, \ldots$, where $q_2 = (1 - p_2)$ and c is a constant chosen to make the total probability one[39].

Let us denote by $S_{x,i}$ the progress in simulation clock due to the processing of the i-th event at LP_x. The values of p_1's and p_2's can be different for different LPs. To maintain notational simplicity, the same p_1 and p_2 values are used in our analysis. The results remain the same under models where they are different. In the sequel, p_1 and p_2 values refer to the probability of success in forward and backward jumps respectively.

• Physical processor time is equally shared among all of the LPs mapped onto that processor. Physical service time $O_{x,i}$ of the i-th event at LP_x is exponential with mean μ_x and $O_{x,i}$'s have independent and identical distribution for all i as long as x remains unchanged.

• Computation of GVT and BVT is performed periodically after every T seconds, and at least one event is processed at every LP during this time period. Process transfer takes place if $BVT - GVT > \Delta$. The distribution function of the random variable $BVT - GVT$ is denoted as $\mathcal{DF}(.)$.

Suppose we start counting time immediately after one GVT computation is over, and one process transfer has taken place between the two physical processors having the fastest and the slowest LPs. We state the following theorem without proof[12].

Theorem 7. 0.4.1 *The process migration scheme is profitable if*
$$\sum_{i=2}^{n_1}\left(\frac{1}{n_1\mu_i} - \frac{1}{(n_1-1)\mu_i}\right) + \left(\frac{1}{n_1\mu_1} - \frac{1}{(n_j+1)\mu_1}\right) + \sum_{i=n-n_j+1}^{n}\left(\frac{1}{n_p\mu_i} - \frac{1}{(n_j+1)\mu_i}\right) > \frac{\Psi}{aRT(1-\mathcal{DF}(\Delta))}, \quad (2)$$
where n_1 and n_j are the number of LPs mapped onto processors $Proc_1$ and $Proc_j$ respectively.

If the above inequality is not satisfied, then there are two controlling parameters, Δ and T. A decrease in Δ implies a decrease in the right hand side of expression (2) . But this cannot be done beyond a certain limit. Indeed, when Δ decreases, the expected cost of load balancing and the total cost Ψ of migrations also increase. An increase in T also decreases the right hand side of expression (2). However, for a very large value of T, the LPs may achieve different simulation clock increments during this interval.

The rate of progress of simulation is measured by the progress in simulation time per unit of physical time. This is an important quantity as it measures the success of the simulation system. Computing the rate of progress of simulation is not an easy task, since it depends on the application problem. We compute bounds for the rate of progress of simulation, modeling the system as a random walk. Suppose our system has run for the physical time interval $[0,T]$. During this interval, each LP_x has processed α_x events such that the time stamp of each of these events follows the geometric distributions specified in our assumption. Hence, the rate of progress of simulation is $Rt = \frac{1}{T}\{Min_{x=1}^{n}\sum_{i=1}^{\alpha_x} S_{x,i}\}$ and the expected rate of progress of simulation is $E(Rt)$.

The following two theorems deal with the bounds on $E(Rt)$.

Theorem 7. 0.4.2 *An upper bound on the expected rate of progress of simulation is* $E(Rt) \leq Min_{x=1}^{n}\frac{1}{\mu_x n_y}((1-a)/p_1 + ac/p_2)$, *where* $y = Map(x)$.

Theorem 7. 0.4.3 *A lower bound on the rate of progress of simulation is given by* $E(Rt) \geq \frac{1}{Max_{x=1}^{n}\{\mu_x\}}E(Min_{x=1}^{n}\{S_{x,1}\})$.

7.4.4.2 Experimental Results

Experiments were conducted[12] to study the theoretical bounds presented in this section, by considering a completely connected system with 8 processors and 48 LPs. The values of μ's were assumed the same for all the LPs. So were for p_1 and p_2. Geometric and exponential random variables were generated, by applying the corresponding inverse transformations on uniform random variables between 0 and 1. The value of $E(Min_{x=1}^{n}\{S_{x,1}\})$ is obtained by iteratively computing the mean of a random variable having the same distribution as $Min_{x=1}^{n}\{S_{x,1}\}$. A successful simulation system, for which these assumptions are valid, should have $p_1 \approx 0$ and $p_2 \approx 1$. This is to ensure that the simulation progresses fast enough in the forward direction. Also if the system is designed well, the probability of rollback should be very low. The authors have also looked at cases varying average service requirements, μ_i's. These experimental results show how the upper,

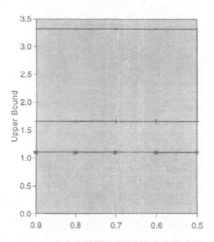

Prob. of success, p2 in backward jump

Prob. of success, p1 in forward jump

— 0.1 + 0.2 * 0.3

Prob. of rollback=0.005, mu=0.5 for all lp's

Fig. 7.5.a Upper bound vs. $p2$

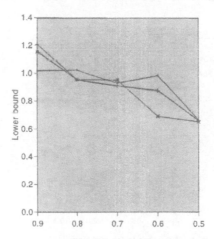

Prob. of success, p2 in backward jump

Prob. of success, p1 in forward jump

— 0.1 + 0.2 * 0.3

Prob. of rollback=0.005, mu=0.5 for all lp's

Fig. 7.5.b Lower bound vs. $p2$

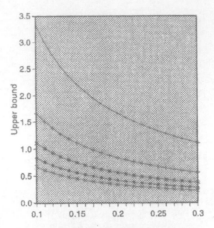

Prob. of success in forward jump

Different values of mu

— 0.5 + 1.0 ✳ 1.5 ◆ 2.0 ✷ 2.5

Prob. of rollback=0.005
Prob. of success in backward jump=0.8

Fig. 7.6.a Upper bound vs. $p1$

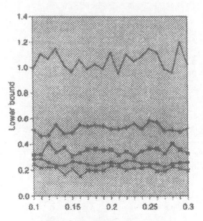

Prob. of success in forward jump

Different values of mu

— 0.5 + 1.0 ✳ 1.5 ◆ 2.0 ✷ 2.5

Prob. of rollback=0.005
Prob. of success in backward jump=0.8

Fig. 7.6.b Lower bound vs. $p1$

and the lower bounds theoretically computed vary with the change in different parameter values. In Fig. 7.5.a the upper bound remains almost constant with the change in $p2$, since the contribution due to this is very less as the probability of rollback itself is very low. But, the lower bound (Fig. 7.5.b) decreases steadily. Figs. 7.6.a and 7.6.b show how the bounds change with μ and $p1$. The upper bound decreases with both μ and $p1$ reaching steady values after a while. Lower bound also decreases with μ but it remains almost constant with changes in $p1$.

7.5. CONCLUSIONS

In this chapter, the problem of load balancing in parallel simulation for execution on multiprocessors is approached in two ways. First, we consider the conservative approach whereby each logical process (LP) is allowed to proceed if and only if it is certain that it will not receive any earlier event. The synchronization protocol makes use of of Chandy-Misra null messages[10]. The use of a simulated annealing algorithm with an adaptive search schedule is described to find good (sub-optimal) partitions. This reported results show that careful static partitioning is important to the Chandy-Misra null message algorithm. We obtained a significant reduction of the run time of the simulations with the use of the simulated annealing algorithm described in this chapter (compared to the use of random placement).

In the second approach, we consider the optimistic approach whereby each LP proceed with its computations without any constraints, and where it is necessary to balance the computational load on different LPs in such a way that their local simulation clocks advance at almost the same rate. A dynamic load balancing algorithm is presented, in which load migration takes place between physical processors. It is based on the global virtual time (GVT) computation. A probabilistic performance analysis of this algorithm on a realistic model, is sketched which derives bounds on the rate of progress of parallel simulation.

We can identify several direction for future research. We plan to continue our studies on the impact of static/dynamic partitioning of conservative parallel simulation[8]. As we have shown in this chapter, dynamic load balancing is a critical issue for exploiting the parallelism in any application, and particularly in parallel simulation where the computational load dynamically changes with both time and space.

Inter-processor communication is a basic factor to contend with on any multiprocessor machine; we would like to evaluate its relationship with the execution time of the simulation in the quest for dynamic load balancing algorithms. We also plan to investigate efficient methods to collect the data necessary for the algorithm and methods to execute the algorithm itself in parallel.

Finally, it is worthwhile to investigate dynamic memory management in parallel simulation with sophisticated flow control schemes.

REFERENCES

1. E.R. BARNES, "An Algorithm for Partitioning the Nodes of a Graph", SIAM J. of Algebraic and Discrete Methods, Vol. 3, 1982, pp. 541-550.

2. H. BAUER, and C. SPORRER, "Reducing Rollback Overhead in Time Warp based Distributed Simulation with Optimized Incremental State Saving", 26th Annual Simulation Symposium.

3. R. BERRENDORF, and J. HELIN, "Evaluating the Basic Performance of the Intel iPSC/860 Parallel Computer ", Concurrency : Practice and Experience, Vol. 4, No. 3, May 1992, pp. 223-240.

4. S. BOKHARI, Assignment Problems in Parallel and Distributed Computing, (Kluwer Academic Publishers), Boston, 1987.

5. A. BOUKERCHE, and C. TROPPER, "A Static Partitioning and Mapping Algorithm for Conservative Parallel Simulation", Proc. of 8th the Workshop on Parallel and Distributed Simulation, 1994, pp. 164-172.

6. A. BOUKERCHE, and C. TROPPER, "An Efficient Distributed Cycle/Knot Detection Algorithm", 9th Int'l Parallel Processing Symp., 1995, pp. 164-172.

7. A. BOUKERCHE and C. TROPPER, "Parallel Simulation On a Hypercube Multiprocessor" Distributed Computing, Vol. 8, pp. 181-190, 1995.

8. A. BOUKERCHE "On Process Migration and Load Balancing In Conservative Parallel Simulation", In Preparation.

9. T. BUI, C. HEIGHAM, C. JONES, and T. LEIGHTON, "Improving the Performance of the Kernighan-Lin and Simulated Annealing Graph

Bisection Algorithms",Proc 26th ACM-IEEE Design Automation Conf. 1989, pp. 775-778.

10. K. M. CHANDY, and J. MISRA,"Distributed Simulation: A Case Study in Design and Verification of Distributed Programs", IEEE TSE-5, 1979, pp. 440-452.

11. J. COHOON, and W. PARIS, "Genetic Placement", IEEE T-CAD, 1987.

12. S. K. DAS, and F. SARKAR,"Reducing Rollbacks Through Load Sharing in Parallel Discrete Event Simulation", Proc. of the Int'l Conf. On Parallel and Distributed Computing and Systems, Oct. 1994, pp. 402-410.

13. S. K. DAS, and F. SARKAR, "A Hypercube Algorithm for GVT Computation and Its Applications in Optimistic Parallel Simulation" , Proc. of the 28th Annual Simulation Symposium, April 1995, pp. 51-60.

14. K. EL-KHATIB, and C. TROPPER, "Dynamic Load Balan cing and Time Warp", TR-SOCS 96, McGill University, Montreal, Canada

15. C.M. FIDUCCIA, and R.M. MATTHEYSES, "A Linear-Time Heuristic for Improving Network Partitions", Proc. of the 19th DAC., 1982, pp. 175-181.

16. R. M. FUJIMOTO,"Performance Measurements of Distributed Simulation Strategies", Proc. of the Workshop on Parallel and Distri buted Simulation, Vol. 19, No. 3, Feb. 1988, pp. 14-20.

17. R. M. FUJIMOTO,"Parallel Discrete Event Simulation",CACM, 33(10), 1990, pp. 30-53.

18. M. R. GAREY, and D. S. JOHNSON, Computers and Intractability: A Guide to the Theory of NP- Completeness, (W.H. Freeman and Company, NY, 1979).

19. D. GLAZER, and C. TROPPER,"On Process Migration and Load Balancing In Time Warp", IEEE Trans. on Parallel and Distributed Sys, tems, Vol. 4, No. 3, March 1993, pp. 318-327.

20. V.B. GYLYS, and J., A., EDWARDS, "Optimal Partitioning of Workloads for Distributed Systems", Proc. COMPCON, 1976.

21. A. HERVE, and C. TROPPER , " Clustered Time Warp and Logic Simulation" Proc. of the 9th Workshop On Parallel and Distri buted Simulation, pp. 112-119, 1995.

22. A. HERVE, and TROPPER, C.,"The Dynamic Load Balancing of Clustered Time Warp for Logic Simulation",Proc. of the 10th Workshop On Parallel and Distributed Simulation, to appear, May 1996.

23. D.J. JEFFERSON, "Virtual Time",ACM TOPLAS. 77(3), 1985, pp. 402-425.

24. D. JOHNSON, C. ARAGON, L. MCGEOCH, and C. SCHEVEON, "Optimization by simulated annealing: An Experimental Evaluation; Part 1, Graph Partitioning", Operations Research, Vol. 37, No. 6, 1989, pp. 865-892.

25. B.W. KERNIGHAN, and S. LIN, "An Efficient Heuristic Procedure for Partitioning Graphs". The Bell System Technical Journal, 49(2), 1970, pp. 291-302.

26. L. KLEINROCK, and S. NILSON ., "On Optimal Scheduling Algorithms for Time-Shared Systems" J. Assoc. Compt. Vol. 28, No. 3, july 1981, pp. 477-486.

27. R. KLING, and P. BANERJEE, "Optimization by Simulated Evolution with Applications to Standard Cell Placement", 27th ACM-IEEE, DAC., 1990.

28. S. KIM, and J. BROWN,"A General Approach to Mapping of Parallel Computation upon Multiprocessor Architectures", Proc. of the Int. Conf. Parallel Processing, 1988, pp. 1-8.

29. S. KIRKPATRICK, C.D. GELATT, and M.P. VECCHI, "Optimization by Simulated Annealing". Science. 220(4598), 1983, pp. 671-680.

30. S. A., KRAVITS, and R. RUTENBAR, "Placement by Simulated Annealing on a Multiprocessor", IEEE Trans. Compt. Design, Vol. 6, 1987, pp. 534-549.

31. B. KRISHNAMURTHY, "An Improved Mincut Algorithm for Partitioning VLSI Networks", IEEE Trans. on Computers, Vol. C-33, May 1984, pp. 438-446.

32. Y. B. LIN, and E.D. LAZOWSKA,"Conservative Parallel Simulation for Systems with no Lookahead Prediction", Tech. TR 89-02-02, Dept. Comp. Sc.& Eng., Univ. of Washington., 1989.

33. B.D. LUBACHEVSKY, B.D. SHWARTZ, and A. WEISS,"Rollback Sometimes Works... If Filtered", Proc. of 1989 Winter Simulation Conf, pp. 630-639.

34. J. MISRA,"Distributed Discrete-event Simulation", ACM Computing

Surveys, Vol. 18, No. 1, March 1986, pp. 39-65.

35. B. NANDY, and W.M. LOUCKS, "An Algorithm for Partitioning and Mapping Conservative Parallel Simulation onto Multicomputer ", Proc. of the Workshop on Parallel and Distributed Simulation, 1992, pp. 139-146.

36. L. NI, et. al., "A Distributed Drafting Algorithm for Load Balancing", IEEE Trans. Software Engineering, Vol. 11, No. 10, 1985.

37. D. M. NICOL, and P.F. REYNOLDS, "The automated Partitioning of Simulations for Parallel Execution", TR-85-15, Univ. Virginia, 1985.

38. D. M. NICOL, "Parallel Discrete-Event Simulation of FCFS Stochastic Queuing Networks", Proc. of the ACM SIGPLAN Symposium on Parallel Programming, Environments, Applications, and Languages, Yale University, July 1988.

39. D. M. NICOL, "Dynamic Remapping of Parallel Time-Stepped Simulations",Proc. of the Workshop on Parallel and Distributed Simulation, Vol. 21, No. 2, March 1989.

40. D. M. NICOL and D. R. O'Hallaron, "Improved Algorithms for Mapping Parallel and Pipelined Computations", IEEE TC-40, No. 3, 1991, pp. 295-306.

41. P. REIHER, and D. JEFFERSON, "Virtual Time Based Dynamic Load Management in the Time Warp Operating System" , Proc. of the Workshop on Parallel and Distributed Simulation, 1990, pp. 103-111.

42. S. M. ROSS, Stochastic Processes, (J. Wiley and Sons, N.Y. 1983).

43. J.H. SALTZ, V.K. NAIK, and D.M. NICOL, "Reduction of the effects of the Communication delays in Scientific algorithms on Message Passing MIMD Architectures", SIAM J. of Sc. and Stat. Compt., Vol. 8, No. 1, 1987, pp. 118-134.

44. L. A. SANCHIS, "Multi-Way Partitioning Network Partitioning", IEEE Trans. on Computers, Vol. 38, No. 1, Jan. 1989, pp. 62-81.

45. F. SARKAR, "Rollback Reduction Techniques Through Load Balancing in Optimistic Parallel Discrete Event Simulation", Ph.D Thesis, Dept. of Compt. Science., Univ. of North of Texas, 1995.

46. V. SARKAR,Partitioning and Scheduling Parallel Programs for Execution on Multiprocessors, (MIT Press, 1989).

47. C. SPORRER, and H. BAUER, "Corolla Partitioning for Distributed Logic Simulation of VLSI-Circuits", Proc. of the Workshop on Parallel and Distributed Simulation, Vol. 23, No. 1, 1993, pp. 85-92.

48. J. STANKOVIC,and I. Sidhu, "An Adaptive Bidding Algorithm for Processes, Clusters and Distributed Groups", Proc. 4th Int'l. Conference Distributed and Comput. Systems, 1984.

49. L. TAO, L., B. NARAHARI, and Y.C. ZAHO, "Partitioning Problems in Heterogeneous Computing", Workshop on Heterogeneous Processing, 1993, pp. 23-28.

50. C. TROPPER, and A. BOUKERCHE, "Parallelizing the Sequencing Problem" IEEE Workshop on Parallel and Distributed Real Time Systems, Newport Beach, Calif. April 1993, pp. 240-244.

51. E.E. WITTE, R.D. CHAMBERLAIN, and M.A. FRANKLIN, "Task Assignment by Parallel Simulated Annealing", IEEE Transaction on Parallel and Distributed Systems, 1991.

52. D.F. WONG, and C.L. LIU,"A New Algorithm Floor Plan Design", Proc. 23rd, DAC Las Vegas, June 1986, pp. 101-102.

53. M.Y. WU, and D. GAJSKI, "A Programming Aid for Hypercube Architecture", Journal of Supercomputing, Vol. 2, 1988, pp. 349-372.

CHAPTER 8

AN OBJECT ORIENTED ENVIRONMENT FOR PARALLEL DISCRETE-EVENT SIMULATION

Rajive Bagrodia and Jerry Waldorf

8.1. INTRODUCTION

Simulation, and discrete-event simulations in particular, have been among the most challenging applications of digital computers. Early language design efforts like SIMULA were motivated, at least partly, by a desire to provide a structured and modular software environment to aid the design of complex simulations. Parallelism has added yet another axis of complexity to the task of simulation model development. Although one would hope that the parallelism in the execution of a model could be made essentially transparent to the analyst, this is not yet feasible for many reasons: a variety of simulation algorithms and techniques have been devised (see Fujimoto[1] for a recent survey), and many of them for the simulation of specific applications. A significant number of performance studies have also been reported on the utility of various suggested techniques in improving the elapsed time of different simulation applications. In most cases where substantial speedups were reported, they were obtained only after the arduous work of low-level performance tuning that optimized specific features of the particular synchronization algorithm. Making the parallelism transparent to the user, without at least committing a priori to a specific protocol, thus seems hard.

165

Conversely, experience with parallel simulators suggests that reduction in the completion time of a simulation depends significantly on the application as well as the specific algorithm used to execute the model on a parallel architecture. For some models, multiple independent replications or even sequential implementations may be more suitable than parallel implementations. There is no existing body of analytical or empirical knowledge that could be used by an analyst to identify the most suitable synchronization protocol for a given application. In the absence of such a priori knowledge about the suitability of a specific simulation algorithm for a given application, it is desirable to design a parallel simulation environment that cleanly separates the issue of model design from that of model execution. This chapter describes an object-oriented simulation environment called MOOSE that allows a simulation model to be elaborated iteratively from an initial prototype to an efficient, possibly parallel, implementation.

In the object oriented paradigm, a program is designed as a collection of message-communicating objects. An object is an abstract data type that defines a set of data structures and a set of operations (commonly called *methods*) that operate on the data structures. Each object is a specific instance of a *class* and a class can be defined in terms of previously defined classes using *inheritance*. The object-oriented paradigm has been successful partially due to its support for modular and reusable software and also for its support of iterative design using stepwise refinements. The simulation environment described in this chapter encourages a programmer to develop parallel simulations using step-wise refinement.

MOOSE (*M*aisie-based *O*bject-*O*riented *S*imulation *E*nvironment) was designed by adding support for inheritance to Maisie, an existing object-based language for discrete-event simulations. Maisie maintains a clear separation between the simulation program and the specific algorithm that is used to execute the program on a parallel architecture. With minor modifications, Maisie programs may be executed using an algorithm based on null messages,[2] conditional events[1] or optimistic space-time computation.[3] A MOOSE pre-processor has been written that translates a MOOSE program into Maisie, thus making all the parallel implementations supported for Maisie directly accessible to a MOOSE program. The Maisie constructs encourage a programmer to develop modular simulations using step-wise refinement, where the refinements progressively transform a prototype to an efficient (parallel) implementation.

MOOSE shows how inheritance can be used to effectively manage the software complexity associated with designing parallel simulations. The design process is separated into two parts:

- design of a validated program which models the physical system at the appropriate level of detail

- efficient implementation of the validated model on a specific (sequential or parallel) architecture.

The purpose of the initial prototype is to ensure that the simulation program models the physical system at the desired level of complexity. As it is generally hard to a priori determine the desired level of complexity, it is useful to initially develop a 'look-and-feel' prototype. MOOSE uses complex guards to specify dynamic enabling conditions for the methods of an object. This allows events and their corresponding enabling conditions to be specified at a high level of abstraction and leads to concise programs. The prototype can subsequently be refined to elaborate the enabling conditions of these events in terms of simpler constructs to improve efficiency. At the prototype stage, the program is executed using a sequential simulation algorithm. If the completion time of the refined sequential program is not acceptable, parallel implementations may be explored.

In the context of simulation models, object oriented design techniques can be used to transform a prototype to an efficient and possibly parallel model with minimal effort. Once the initial prototype has been validated, specific parts of the model are elaborated by using inheritance to redefine some of the attributes of the object and provide additional functionality. Once a model of the physical system has been designed at the appropriate level of detail, optimizations to improve the execution efficiency of the model may also be added through inheritance. The optimizations can appear in various forms: for instance if the model is executed on a parallel architecture using an optimistic simulation algorithm, an object, say $e1$, can be refined using inheritance to define another object $e1\text{-}optimistic$ which may improve its performance either by reducing the state information that needs to be saved or by generating conditional future messages as early as possible in its execution. Subsequently, if the model is executed using a conservative parallel algorithm, the object $e1$ may be refined to improve its lookahead characteristics and thus improve its performance. Alternately, if the model is moved to a different parallel architecture, the object may be refined to better exploit the architectural features to improve its efficiency. Note that the use of inheritance allows the original simulation model to be separated cleanly from its optimized versions.

The remainder of this chapter describes the MOOSE environment. The next section describes related work — we compare our approach to other parallel object-oriented languages and also to object-oriented simulation languages. Section 8.3 gives an overview of the primary constructs of the MOOSE language. Section 8.4 gives a detailed description of the synchronization and inheritance facilities provided by MOOSE. Section 8.6 describes the use of inheritance in simulation development. Section 8.5 illustrates the use of MOOSE in designing parallel implementations of simulation models using a closed queuing network. Section 8.8 contains a brief description of the implementation of MOOSE. Section 8.9 is the conclusion.

8.2. PARALLEL OBJECT-ORIENTED LANGUAGES

In this section, we outline a set of four requirements that must typically be satisfied by parallel object-oriented languages. Existing languages are evaluated within the framework of these requirements; the section concludes with a brief look at how MOOSE meets the stated requirements; in particular, we indicate how the inheritance mechanisms provided by MOOSE differ from those in other parallel object-oriented languages.

8.2.1. Requirements

In many ways, parallelism and object-orientedness are orthogonal concerns in the design of a programming language; this partially explains the large number of proposals that have been suggested in the literature for the design of parallel object-oriented languages.[4] In an effort to isolate the key design issues, we identify some characteristic features of parallel object-oriented languages. These characteristics are used to compare the many proposals to identify their major strengths and weaknesses. The primary features provided by parallel object-oriented languages include mechanisms to support data encapsulation, object synchronization, concurrent execution and inheritance.

Encapsulation refers to the ability of a language to restrict access to the data structures defined within an object, such that this access is provided only via well-defined interfaces called *methods*. A specific method may be invoked by sending an appropriate message to the object.

Synchronization refers to the ability of an object to specify the subset of its methods that may be executed in any given state. The state of an object is determined entirely by the value of its local data structures. In a given state of the object, a method is said to be *enabled* if it can be executed; otherwise the method is said to be *disabled*. For instance, consider a stack object. If the stack is empty, the *pop* method should be disabled. We use the term *enabled set* to refer to the subset of methods defined for an object that are enabled in a given state of the object. The enabled set for an object may either be specified using only its local state, or it may also depend on the contents of its message buffer. We refer to the former as *static* synchronization as the enabled set of a suspended object is static, and the latter as *dynamic* synchronization as the enabled set may be changed by the arrival of additional messages for the object.

The basic form of *concurrency* in an Object-Oriented Language (OOL) is provided by supporting simultaneous execution of enabled methods in different objects. Additionally, some languages may also support internal concurrency, by allowing simultaneous execution of multiple methods defined within a single object.

Lastly, *inheritance* refers to the ability to define a new class of objects in terms of existing object classes. The derived class (or equivalently object) must be able to selectively inherit the data, methods and synchronization defined in one or more base classes.

8.2.2. Related Work

We now examine a number of parallel OOLs with respect to the criteria defined above. We consider both general purpose parallel OOLs like POOL,[5] ACT++,[6] and Rosette[7] as well as OOLs that were specifically designed to support parallel execution of simulation programs as typified by MODSIM II[8] and Sim++.[9] All the preceding languages provide similar mechanisms for encapsulation. We compare the languages with respect to the other criteria.

Synchronization. MODSIM II does not have any provisions to define enabled sets; in other words, all methods defined by an object are assumed to be enabled at all times. POOL and Sim++ adopt similar approaches to synchronization: Each POOL (or Sim++) object defines a set of methods and a *body*. The body must be executed after invocation of every local method to determine the enabled set. Whereas a POOL object may only use local data to define the enabled set, a Sim++ object may also reference message parameters. As seen subsequently, the use of a centralized body to determine the enabled set makes it very hard for derived objects in these languages to inherit synchronization information. Instead of using a central location, ACT++ and Rosette distribute the code required to compute the enabled set of an object following the execution of a specific method.

Concurrency. MODSIM II supports concurrent execution by providing both synchronous and asynchronous method invocation. Both Sim++ and POOL provide a body that is a local process. The body executes in parallel with other objects sending and receiving messages when necessary. POOL uses synchronous message passing whereas Sim++ uses asynchronous communication primitives. ACT++ and Rosette objects can execute concurrently and communicate via asynchronous message passing.

Inheritance. MODSIM II provides inheritance of methods and data but since there is no notion of conditional synchronization in the language, synchronization cannot be inherited. Sim++ and POOL do not allow inheritance of synchronization. In POOL one can only inherit the functions defined in the object, the body must be redefined for each derived object. Sim++ provides the simulation entity to the programmer as a base class that must be inherited by every entity that performs typical simulation tasks that include being delayed or suspended in simulation time. When deriving an entity from another, inheritance is limited as in POOL such that only the data or methods defined in the base class can be inherited. In ACT++ and Rosette, a derived object can inherit not only the methods and data but also the synchronization mechanism, which however is static.

MOOSE. In POOL all of the synchronization information was placed in the body. In the actor based languages, the information about enabled methods was specified in a set. Unlike the preceding languages, the synchronization information in a MOOSE object is specified by including a Boolean condition or *guard* with each method. Taken together, the guards on all the methods define the enabled set of the object. Distributing the synchronization information in this manner has two primary advantages: first, it facilitates inheritance of the synchronization information and second, when adding new methods in a derived class, it is only necessary to define the guard for the new methods; the guards on the inherited methods do not have to be modified. In particular, MOOSE is the first object-oriented simulation language that supports dynamic synchronization and also allows the synchronization information to be inherited separately from the methods defined for an object.

8.3. MOOSE OVERVIEW

A MOOSE program defines a collection of entity-types. An entity-type is similar to a class and represents objects of a given type. An entity-instance, henceforth referred to simply as an entity, represents a specific object of the corresponding class and may be created and destroyed dynamically.

Each entity-type defines a set of data structures and a set of *methods* that operate on the data. A method definition is similar to a function in that it has a name, a set of formal parameters and a body. A method may additionally be associated with a *guard*, a Boolean expression that may reference the method parameters and local data. A method may be executed only if the value of its guard is true. Execution of the method proceeds by executing the statements specified in its body. Two special methods, **create** and **destroy**, may be defined for each entity-type. These methods are respectively executed when an entity is created or destroyed. For instance, consider the *manager* entity-type defined in Figure 8.1. This entity-type models a resource manager. The entity-type has a single data item called *units* and three methods respectively called *req* (line 14-17), *free* (lines 18-20) and *create* (line 21). When a manager entity is created, the specified *create* method is executed to initialize the entity with a given number of resource units. Subsequently, the *req* method is used to allocate a resource and *free* is used to return a resource to the manager. Although the *free* and *create* methods do not need a guard, the guard on the *req* method ensures that it is executed only if resource units are available with the manager.

```
1   entity driver {
2       ename r1;
3       create() {
4           r1 = new manager(10);
5           ...
6           invoke r1 with req(self);
7       }
8       done() {
9           hold(t);
10      }
11  }
12  entity manager {
13      int units;
14      req(ename id) st(units > 0) {
15          units − −;
16          invoke id with done;
17      }
18      free() {
19          units + +;
20      }
21      create(int num) { units = num; }
22  }
```

Figure 8.1 A Resource Manager

An entity is created by the execution of a **new** statement. When created, every entity is assigned a unique identifier. An entity can reference its own identifier using the keyword **self**. MOOSE also defines a type called **ename** which is used to store entity-identifiers. Figure 8.1 includes a *driver* entity that creates a *manager* entity (line 4). Execution of this statement creates an instance of entity *manager* with 10 *units* and stores its identifier in variable r1. (Note that variable r1 must be of type **ename**). Every MOOSE program must have a *driver* entity. This entity initiates execution of the simulation program and serves essentially the same purpose as the main function in C. MOOSE also provides a **delete** statement; execution of this statement destroys the specified entity.

We now consider method execution. An entity, say *e*1, executes a method in another entity (the two entities need not be distinct) by executing an **invoke** statement. The invoke statement must specify the id of the destination entity and provide the name and actual parameters for the method being invoked. For instance, in our example, the *driver* entity uses the invoke statement in line 6 to execute the *req* method in entity *manager*. As a method may be associated with a guard, it follows that a request of an entity to execute a remote method in another entity may not be honored immediately by the destination entity; rather than blocking the invoking entity until its request can be honored, the request is simply buffered in the destination entity and the invoking entity resumes its execution. Thus execution of an invoke statement simply deposits a message in a message buffer associated with the destination entity; the message is removed from the buffer only when the destination entity executes a corresponding method.

A method, say *m*, defined in entity *e* is said to be *enabled* if the message buffer of entity *e* contains a message of type *m* and the guard defined for the method evaluates to true, (if the guard is omitted, it is assumed to always have the value *true*) the corresponding message is referred to as an *enabling message*. A method that is not enabled is said to be *disabled*. For instance, the *req* method defined in entity *manager* will be enabled if the message buffer includes a *req* message and if *units* is greater than 0 (i.e., resource units are currently available with the manager). The *free* method is enabled whenever the message buffer contains a *free* message. If two or more methods defined in an entity are enabled, one of them must be selected for execution using a deterministic criterion. Every message is assigned a timestamp (the algorithm to assign message timestamps is described subsequently). If two or more methods are enabled, the timestamps on the corresponding enabling messages are compared and the message with the earliest timestamp is removed, delivered to the entity and the corresponding method is executed. If the earliest timestamp is not unique, any one of the corresponding enabling messages may be delivered.

Execution of a method proceeds by simply executing the statements in its body. For instance, execution of the *req* method in entity *manager* (line 15-16) allocates the resource to the requesting entity by means of an invoke statement. Similarly, execution of the *free* method (line 18) adds the returned unit to the available resource pool.

In general, the enabling condition for a method may include multiple message-types, each with its own guard. This allows many complex enabling conditions to be expressed directly, without requiring the programmer to describe the buffering explicitly. Enabling conditions in their most general form are discussed in the next section.

```
1     entity newmanager
2       inherit manager
3     {
4       multreq(ename id, int count) st(units > count) {
5         hold(exp(count));
6         units- = count;
7         invoke id with done;
8       }
9       req(ename id) st(units > 0) {
10        hold(exp(1));
11        units - -;
12        invoke id with done;
13      }
14    }
```

Figure 8.2 A Derived Resource Manager

We now give an overview of the type of inheritance supported by MOOSE; the next section provides a detailed description. Inheritance may be used to define a new entity-type by adding additional methods to an existing entity-type, or it may be used to derive a new entity-type by modifying the definition of one or more methods, where the modifications may involve either the body or the guard associated with the method. MOOSE does not currently support multiple inheritance.[10,11] The newmanager entity-type defined in Figure 8.2 illustrates both types of inheritance. Line 2 indicates that the new entity-type inherits the data structures and methods defined in the base entity-type manager.

The method multreq is an example of an addition to the entity (lines 4-8). Multreq is added to the set of methods inherited from manager. Unlike the req method, this method allows an entity to simultaneously request multiple units of the resource. The multreq method defines two parameters, *id* and *count*, where the second parameter refers to the number of units requested.

Redefinition of an inherited method is demonstrated by the new definition of *req* in lines 9-13. This method overrides the *req* method defined in *manager*. Method *req* now includes a call to hold (line 10) which simulates the processing time required to find an available resource and return it to the requester.

We now consider the MOOSE constructs that are useful for writing discrete-event simulations. A discrete-event simulation is a sequence of events, where each event simulates some activity of interest in the physical system and may involve one or more objects. In developing a model of the resource manager, events include 'a job requesting a resource' or 'a job using the resource for t time units'. Events in a MOOSE model are simulated by method invocations. For instance, the first event is modeled by a job entity sending a message to execute the *req* method in the *manager* entity. Events that involve the passage of time must be modeled separately as described subsequently.

Every MOOSE program is associated with a *clock* variable that represents the simulation clock. Every message generated in the simulation is implicitly time-stamped with the current value of the simulation clock. Although the value of the clock variable cannot be modified directly by any other object, a set of methods that operate on this variable are provided by a system-defined object called *simset*. These methods are defined as functions and can be called directly by any entity that is derived from this entity.

The *simset* entity defines three primary functions: a parameterless function *sclock*() that returns the current value of the simulation clock, and functions *hold* and *timeout* that are used by an entity to simulate the passage of physical time in the system. Function $hold(t)$ is executed by an entity to suspend itself unconditionally for t units of time. This function may be used by an entity to simulate the passage of time in performing an unconditional event like 'using the resource for t time units'. However, a simulation may also include conditional or interruptible events; a simple example is an event like the service of a low priority request by a priority preemptible server. This event is said to be conditional because it can be interrupted by the arrival of a high priority request. An interruptible event can be scheduled using the *timeout* function. This function has two formal parameters: t which is an integer that represents a time-interval and *action* whose type is a function pointer. If an entity executes *timeout(action, t)*, it ensures that function *action* will be executed, if no enabling message is available in the message-buffer of the entity within *t* time units.

```
1    entity server
2        inherit simset {
3        int hcnt = 0, lcnt = 0, remtim, lostart, state = IDLE;
4        int cmeanh, int cmeanl;
5        create(int cmh, int cml) {
6            cmeanh = cmh; cmeanl = cml;
7        }
8        high() st (state == IDLE) {
9            hold(exp(cmeanh)); hcnt + +;
10       }
11       low() st(state == IDLE) {
12           state = LOW; lostart = sclock(); remtim = exp(cmeanl);
13           timeout(lowfinish, remtim);
14       }
15       high() st (state == LOW) {
16           remtim − = sclock() − lostart;
17           hold(exp(cmeanh)); hcnt + +;
18           timeout(lowfinish, remtim);
19       }
20       local lowfinish(){
21           state = IDLE; remtim = 0; lcnt + +;}}
```

Figure 8.3 MOOSE Model of Priority Server

As a simple example, consider the simulation of a preemptible priority server in MOOSE. In the physical system, the server receives two types of requests, respectively referred to as *high* and *low*, where the requests of the first type have a higher priority and can interrupt the server if it is currently serving a request of type *low*.

Figures 8.3 and 8.4 describe the MOOSE model of the system. In the interest of brevity, the program ignores issues concerned with simulation initiation and termination, which are discussed in Section 8.5.

Entity-type *server* models the priority server and *hisrc* and *losrc* respectively model the sources for the two types of requests. The *server* entity defines two types of messages, *high* and *low*, to represent the two

```
1    entity hisrc
2      inherit simset
3    {
4      create(ename srvrid; int meanh;) {
5        while (true) {
6          hold(exp(meanh));
7          invoke srvrid with high;
8        }
9      }
10   }
11   entity losrc
12     inherit simulation-entity
13   {
14     create(ename srvrid; int meanl;) {
15       while (true) {
16         hold(exp(meanl));
17         invoke srvrid with low;
18       }
19     }
20   }
```

Figure 8.4 MOOSE Model of High and Low Source

types of requests that may be received by it. Henceforth, we will use *high* message to mean a message of type *high*; similarly for *low*. When *IDLE*, the entity accepts the next message from the buffer, whether it is a *high* or *low* message. If the idle entity accepts a *high* message (line 9), it executes a hold statement (line 10) to simulate the servicing of the request. When completed it simply increments the count of requests that have been serviced (line 10). If the entity accepts a *low* message, it conditionally schedules its completion by using the time-out function (line 14). If no *high* message is received in the interval *remtim*, the entity will execute *lowfinish*, specified as the time-out function. *Lowfinish* is defined as a **local** function. A local function is only accessible from within the entity. If a *high* message is received while the server is servicing a *low* message (line 16), the *high* message is processed by executing the hold function, after which the service of the *low* message is resumed by again executing the time-out function (line 19). When function *low finish* is finally executed, the number of *low* jobs is incremented (line 22).

The two source entities (Figure 8.4) simply generate requests with the appropriate priority at periodic intervals sampled from an exponential distribution. The hold function in each source (line 6 and 16) is used to delay the entity by the corresponding interarrival time for each request type.

8.4. MOOSE CONSTRUCTS

The two main extensions added by MOOSE to Maisie are the method construct and inheritance. In this section we will discuss each of these in detail.

8.4.1. Methods

In the examples described in the previous section, a method was typically defined together with its guard and its body. Furthermore, a method and its parameters implicitly defined a message-type; messages of this type were used to execute the corresponding method. The enabling condition for a method was thus defined on the basis of a single message-type. In this section, we first describe the separate notion of a *message-type* and subsequently show how it is used to define a method.

MOOSE defines a type called **message**, which may be used to explicitly define the types of messages that may be received by an entity. The definition of a message-type is similar to that of a struct. The type definition begins with the keyword **message** followed by a name for the type and an optional list of typed parameters (or fields). For instance, the following fragment defines a message called *push* with a single integer parameter called *item*.

> **message** *push*(int *item*)

However, as long as each message can activate exactly one method, the message-type can be defined implicitly by the method definition and need not be defined separately in the entity. We first consider the simpler types of method definitions where the message-type is defined implicitly. The simplest form of a method definition is illustrated by the following entity, where the *push* method implicitly defines the message-type *push* described previously. This method will be enabled if the message buffer contains a message of type *push*. The actions to be executed by the method are specified in the body.

```
entity stack-obj {
    int stack[100], cnt;
    ...
    push(int item) {
        stack[cnt + +] = item;
    }
}
```

In the above example the method *push* is unguarded, that is, the method is enabled (and can hence be executed) whenever a *push* message is available. In particular, the method can be executed even if the stack is full, which implies that the error-handling or temporary buffering must be done explicitly within the method. In contrast, it may be desirable to ensure that the method is enabled only under restricted conditions that are determined by the current state of the entity, the contents of an incoming message, or both. For instance, in the preceding case, it may be desirable to enable the *push* method only if the stack is not full. To allow this facility, a method definition may optionally include a Boolean expression or guard indicated by the keyword **st** (mnemonic for such that) as illustrated by the following modified definition of method *push*:

```
push(int item) st (cnt < size) {
    stack[cnt + +] = item;
}
```

The guard associated with a method may also refer to message parameters. In the next example, the method is enabled only if the stack is not full and if the parameter *item* is positive:

```
push(int item) st (cnt < size ∧ item > 0) {
    stack[cnt + +] = item;
}
```

We now consider the definition of an entity where the message and method declarations are independent. Consider the situation where rather than delaying execution of *push* until the stack becomes non-full, a separate method is defined to handle this case. In the following example, we first declare the message-type *push* and include an additional parameter *sender* of type ename. Subsequently we define two methods that correspond to this message: the first one is enabled whenever the stack is not full, and the second one is enabled if a *push* message is received when the stack is full. In the second case, the method body sends an appropriate error message.

```
message push(int item, ename sender)
push st (cnt < size) {
    stack[cnt + +] = item;
}
push st (cnt ≥ size) {
    invoke sender with error;
}
```

In the preceding example, the guards on the two methods are mutually exclusive such that at most one guard is enabled at any time. In general, overlapping guards may be defined for the methods, such that two or more methods are simultaneously enabled for the same message. In this case, the choice of the specific method to be executed is defined to be non-deterministic (in other words, the choice of a specific method for execution is dependent on the implementation of the specific compiler).

8.4.1.1. Method body

In most of the preceding cases, the body of a method is defined as a sequence of statements, together with the method heading. In general, the body may simply call a local function and two or more methods may use a common function, possibly with different parameters. This is illustrated by the following example:

```
alloc(int size) {
    ...
}
malloc(int size) alloc(size)
```

calloc(int *size*, int *number*) *alloc*(*size* * *number*)

8.4.1.2. Multiple messages

In its most general form, the enabling condition for a method may include multiple messages (possibly of different types). The last example describes this variation.

As long as the enabling condition of a method referenced a single message-type, the name of the message-type could also be used to name the method. However, if multiple messages are used in its enabling condition, a unique method name must be specified explicitly as illustrated by the next example. This fragment defines a method called *both*, whose enabling condition consists of a *pop* message and a *push* message that contains a positive value. The method body simply returns the value being pushed to the entity that sent the *pop* message. Note that the entity may define additional methods that use the *push* and *pop* message-types.

```
both :  pop(ename req) ∧ push(int num) st (num > 0) {
    invoke req with ret(num);
}
```

8.4.2. Inheritance

We now describe the use of inheritance to derive a new entity from an existing one. Once again, the description is example driven; a complete syntax is given in the appendix.

8.4.2.1. Addition of methods

The simplest form of inheritance allows a new entity to be derived from a base entity by defining additional methods for the entity. Consider an entity called *boundbuf* that models a bounded buffer and defines the standard *put* and *get* methods. The *put* method inserts an item into a buffer that is not full, and a *get* method sends the next item, in *fifo* order, to a requesting entity. It is required to define a new buffer entity-type, which in addition to the *put* and *get* methods also provides a method called *getlast* which operates on a non-empty buffer

to return the last item added to the buffer, rather than the first. This
entity may be derived by inheriting the data structures and the *put* and
get methods defined in entity *boundbuf* and adding the new method
as illustrated by the following fragment. Note that the new method
implicitly adds a new message-type to the entity.

```
entity eboundbuf
  inherit boundbuf
{
    getlast(ename id) {
      ...
    }
}
```

8.4.2.2. Message reuse

It may sometimes be necessary to add additional methods to an entity without introducing a new message-type. This is necessary, for
instance, in the case of a resource manager, where rather than delay
a request that cannot currently be satisfied, the manager immediately
sends a *denial* to the requesting entity. The following example derives a
new resource manager entity from the *manager* entity of the previous
section that defines a method to handle the preceding scenario.

```
entity newmanager
  inherit manager
{
    denyreq : req st(units <= 0) {
      invoke id with denial;
    }
}
```

8.4.2.3. Method redefinition

Another use of inheritance is to change the meaning of an attribute
defined in the base entity. Previously we used inheritance only to add
new functionality to an entity. Now we introduce the powerful notion
of redefinition. In the following example the new resource manager
inherits all of the attributes of its base class except for the *req* method;

this method is redefined in the derived class to accept requests that explicitly indicate the number of units that are being requested. The method heading is modified to introduce an additional parameter *count* that specifies the number of units requested. The modified body is omitted for brevity.

```
entity manager {
  req(ename id) {
    ...
  }
  ...
}
entity newmanager
  inherit manager
{
  req(ename id, int count) st (cnt > count) {
    /* redefinition of heading and body for the req method *
  }
}
```

8.4.2.4. Message redefinition

The preceding example was an illustration of the redefinition of a method. In the same way, we can also allow the redefinition of a message, where the method body remains unchanged. For instance, the following example first describes the functionality of a channel entity; it is subsequently used to derive a specific *channel* that is used to transmit messages of a specific type. This facility is useful in defining library objects, where data types may subsequently be redefined by the user.

```
entity channel {
  ...
  message m()
  rec : m {
    invoke e with m;
  }
}
entity mychannel
  inherit channel
{
```

message m(new message parameters)
}

8.5. EXAMPLE

In this section we develop a MOOSE model for a simple queuing network[12] and subsequently refine it for parallel execution using a conservative algorithm. We also provide completion time measurements for the parallel implementations of the network.

Server Queue   Router

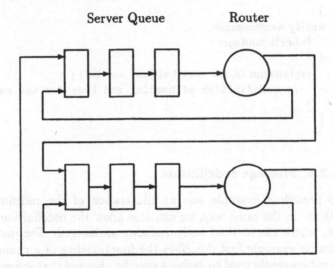

Figure 8.5 Model of a Two Switch CQNF network

Consider a closed queuing network (henceforth referred to as CQNF) that consists of N fully connected switches. Each switch is a tandem queue of Q fifo servers. A job that arrives at a queue is served sequentially by the Q servers and is thereafter routed to one of the N neighboring switches (including itself) with equal probability. The service time of a job at a server is generated from a negative exponential distribution, where all servers are assumed to have an identical mean service time. Each switch is initially assigned J jobs that make a predetermined number of trips through the network.

The MOOSE model of this network consists of two primary entities: a *server* entity to model the tandem servers in a queue and a *router*

```
1    entity server {
2       int mtime, seed; ename creator;
3       int ntime, ename rid;
4       local processjob(int id, int twait, int dep) {
5          twait += sclock() - dep;
6          hold(ntime);
7          ntime = expon(mtime, seed);
8          invoke rid with
9             job(id, twait, sclock());
10      }
11      idmsg(ename id) {
12         rid = id;
13      }
14      job(int id, int twait, int dep) {
15         processjob(id, twait, dep);
16      }
17      create(ename c, int m, int s) {
18         creator = c; mtime = m; seed = s;
19         ntime = expon(mtime, seed);
20      }
21   }
```

Figure 8.6 Server for CQNF

entity that routes a job after it has completed service at a queue. Figure
8.5 displays the model of a network with N=2 and Q=3. Each job in
the network may be modeled as a separate entity or be abstracted by a
sequence of messages. We adopt the latter approach. We will examine
the *router* (Figure 8.7) and *server* (Figure 8.8) entities in detail.

We first consider the *server* entity (Figure 8.8) that simulates service of incoming jobs. When a job arrives the server holds for a time to
simulate servicing the job. The server then calculates the next service
time (line 7). Finally upon completion of service, the job is forwarded
to the next server (line 8-9).

The jobs initially allocated to each switch of the physical network
are allocated to the corresponding *router* (Figure 8.7). On being created, a *router* entity distributes these jobs among the various *server*
entities (line 15). Subsequently, for each incoming job, if the incoming

```
1    entity router {
2      int mtrips;
3      ename qids[N];
4      job(int stime, int twait, int dep) {
5        if(+ + count < mtrips) {
6          invoke qids[iurand(0, N − 1)]
7              with job(stime, twait, dep);
8        }
9      }
10     create(int njobs, int mt, ename qd[N]) {
11       int i;
12       mtrips = mt;
13       for(i = 0;i < N;i + +) qids[i] = qd[i];
14       for(i = 0;i < njobs;i + +)
15         invoke qids[i%N] with job(0, 0, 0);
16     }
17   }
```

Figure 8.7 Router for CQNF

job has not completed its required number of trips, the *router* entity generates a future message that simulates arrival of the job at the next switch in the network. Note that if the incoming job has completed the required number of trips, no additional messages are generated or scheduled. This implies that the event-list becomes empty when each job in the system has completed the required number of trips.

The preceding MOOSE program was executed on a Sun Sparcstation for a network of 16 switches. The configuration information together with the computed statistics are shown in Figure 8.8. The next section describes parallel implementations of the model.

8.6. OPTIMIZATIONS WITH INHERITANCE

The execution efficiency of a simulation model may be improved by executing it on a parallel architecture. Parallel simulation algorithms may be broadly classified into two categories: conservative[2,13] and opti-

```
CQNF Configuration simulated:
        Number of routers = 16
Initial no. of jobs/router = 16
   Number of servers/queue = 10
      Number of trips/job = 10
Statistics Collected: Average System Time
Total number of values: 514
        Mean value: 384.60117
     Maximum value: 946.00000
     Minimum value: 41.00000
```

Figure 8.8 Sample Run of CQNF

mistic.[14,15] Conservative algorithms require that an object process a given message only when it can ensure that it will not subsequently receive a message with a smaller timestamp. In contrast, optimistic algorithms allow an object to process messages out of timestamp order, and use rollback and recomputation to correct timing errors. The primary source of overhead in conservative methods is the periodic synchronization that is needed to identify messages with the globally earliest timestamp. The major overhead in optimistic methods include checkpointing and rollback costs.

The performance of a parallel simulation depends significantly on the ability of the simulator to minimize the overheads that are inherent in the corresponding algorithm. For instance, the performance of conservative simulators may be improved by aggressively exploiting lookahead[16] in the application, whereas that of optimistic implementations may be improved by reducing checkpointing and rollback overheads. Although a class of optimizations may be implemented transpar-

```
1    entity optimizedConservativeServer
2      inherit server, conservative {
3        job(int id, int twait, int dep) {
4          processjob(id, twait, dep);
5          lookahead(ntime); }
6    }
```

Figure 8.9 Parallel CQNF: Derived entity

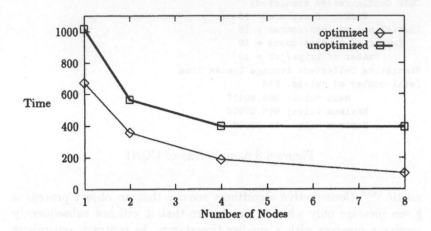

Figure 8.10 Runtimes for CQNF

ently by a simulator,[17] algorithm and application-specific optimizations may be useful in further reducing the execution time of an application. As the suitability of specific optimizations for a given model are not necessarily known a priori, it is desirable to enforce a strict separation between the model and the modifications designed to improve its execution efficiency with a specific algorithm. The inheritance features of MOOSE were designed to enforce this separation and facilitate the introduction of the optimizations in a modular manner.

A detailed discussion of the optimizations is beyond the scope of this paper. Optimizations for optimistic algorithms have been described previously,[17] as well as those for conservative algorithms.[18] In the remainder of this chapter, we describe how some of these optimizations may be used to improve the execution efficiency of a simulation model. To provide an easy way to integrate these optimizations in a model, the MOOSE environment provides two base classes, respectively called *conservative* and *optimistic*. Each of these classes provides a set of data and function members that can be used by a derived class to implement these optimizations. This section examines optimizations with conservative algorithms; optimistic algorithms are considered in the next section.

Precise knowledge of the communication topology of the network and *lookahead* properties of the system are two factors that have a significant impact on the performance of conservative algorithms. Good lookahead is probably the most important determinant for the performance of a conservative implementation. A process is said to have a lookahead of ϵ, if at simulation time t, the process can predict the outputs generated by it in the interval $[t, t + \epsilon]$. The most common use of this attribute is to allow a process to specify a lower bound on the timestamp of any future messages that may be generated by it. This information is used by entities in the destination-set of a given entity to reduce synchronization overheads.[18] The lookahead for an entity typically varies dynamically and is determined by the behavior of the physical process that it models as well as on the message arrival pattern. Although it is sometimes possible to extract the lookahead transparently from a simulation model, it is more effectively expressed directly by the programmer. The base *conservative* class provides a function called *lookahead* which may be called by a derived class to set the earliest (simulation) time at which it will send any future message.

By providing classes that implement the various algorithm specific functions used in the simulation, the programmer can override some of these methods for specific situations in the model. These changes to the underling implementation will provide increases in performance that may not be deducible by a compiler.

Consider a first-in-first-out server whose service time for incoming jobs is sampled from a random distribution. When busy, its lookahead is the remaining service time of the job that is currently in service; when idle, its lookahead is equal to the service time of the next job which can be precomputed by the server. Figure 8.9 illustrates how the server entity of the CQNF example from the previous section can be refined to derive a conservative implementation that exploits lookahead. The program was executed using a conservative simulation algorithm. The experiments reported in this paper were conducted on a Symult 2001 hypercube. Each node of the multicomputer uses a Motorola 68020 CPU and has 4MB memory. The sequential version used for the comparison was executed on a single node of the same machine using a sequential simulation algorithm. Figure 8.10 shows the increase in performance that was obtained using the optimizations described for the *server* entity.

Precise knowledge of the communication topology can also be useful for null message algorithms. In the absence of this information, the null messages would have to be broadcast which could severely degrade performance for sparse networks. For static topologies, the destination set of each entity can be estimated to contain every entity whose id is stored locally as the value of some variable of type *ename*. As the communication topology may change in a dynamic manner, maintaining this information could yield a significant performance advantage. The base *conservative* class provides a set of four functions respectively called *add_source, del_source, add_destination*, and *del_destination* that can be used by the derived entities for this purpose. Sufficient conditions to guarantee correctness of a conservative implementation when used to simulate a dynamic network are described elsewhere.[18]

8.7. OPTIMISTIC OPTIMIZATIONS

A number of optimizations can be defined to improve the performance of an entity with an optimistic simulator. The primary goal of these optimizations is to reduce the state saving and recomputation overheads. We first discuss state saving overheads.

8.7.1. State Saving

The frequency with which the state of an entity is checkpointed together with the size of the state to be checkpointed can have a significant impact on the execution efficiency of an optimistic implementation. Experimental studies have shown that most rollbacks in an optimistic implementation are over fairly short distances. As a result, it is generally most efficient to save the state of an entity before processing every event. Furthermore, by default, all data members of an entity are saved during each checkpoint.

The efficiency of an optimistic implementation may be improved by eliminating redundant checkpointing either where the state being checkpointed is a *dead* state or it has not been modified. A *dead* state refers to a specific checkpointed state, where no recomputation ever begins from that state; such a state need not be copied and saved in the state queue. A time-out message scheduled as a definite event[17] is the most common example of a dead state. As another example of redundant checkpointing, consider an entity that receives a number of queries implemented as methods, where each query simply returns the value of a state variable but does not modify the state of the receiver. Finally, even when the state of an entity is modified, the modifications may be restricted to only a small subset of the state. In each of the preceding cases, checkpointing overheads could be reduced significantly by allowing the programmer to explicitly control the state saving.

The base *optimistic* class contains a function called *save_state*. The default actions associated with this function is to checkpoint all variables of a derived entity prior to the execution of any of its methods. However, a derived entity can redefine this method such that its state is checkpointed only at selected points in its execution; in particular, the function can be redefined such that dead states and states in which a query message is processed are not checkpointed.

```
1    entity server
2      inherit optimistic {
3      int nojobs; ename creator;
4      int idle; ename rid;
5      local processjob(int id, int stim) {
6          idle = false;
7          hold(stim);
8          invoke rid with
9              job(id,stim); idle = true;
10             nojobs = nojobs + 1;
11         }
12     job(int id, int stim) {
13         save_state();
14         processjob(id, stim);
15         }
16     }
```

Figure 8.11 Server

When it is discovered that a method call operates on a deadstate then the state does not need to be saved. The state is saved only under certain conditions as illustrated in Figure 11.

8.7.2. Rollbacks

An optimistic simulation is typically rolled back if the runtime system detects that an out of order message is *delivered* to an entity. We use the term *straggler* message to refer to a message that is delivered to an LP after a message with a larger timestamp has been delivered.

Consider two messages m_a and m_b which are to be delivered to an entity $e1$ at simulation time t_a and t_b respectively, where $t_a < t_b$. Assume that message m_b is delivered first; subsequently when the straggler message m_a is received, entity $e1$ is typically rolled back to a checkpointed state with a timestamp smaller than t_a. The total number of events that must be recomputed when a rollback is initiated is referred to as the rollback distance. The optimizations discussed in this section aim to reduce the rollback distance of an optimistic implementation.

Probe methods.   A *probe* method refers to a method whose processing does not alter the state of the entity. A probe method is typically used to obtain state information about the entity, such as whether it is *active* or *idle*. We use the term *probe* message to refer to a message that executes a probe method. Because a probe message does not need its state saved it need not call the save_state method. The following example illustrates the use of identifying probe methods.

Now that we have identified the probe method we can remove the call to save the state. Note that all other methods defined in the *queue* class will still save state only the *queuelength* method does not.

In the previous section, we showed that it is not necessary to save the sate of an entity prior to processing a *probe* method. We now show that if a straggler message is identified to be a probe, it can be processed without initiating an expensive rollback. The base optimistic

```
1    entity optimizedOptimisticServer
2      inherit server {
3      job(int id, int stim) {
4        if (stim ¿ x) save_state();
5        processjob(id, stim);
6    }}
```

Figure 8.12 Parallel server: Derived entity

```
1    entity queue
2        inherit optimistic; {
3        int qlength;
4        ...
5        quelength(ename id) {
6            save_state();
7            ...
8            invoke id with result(qlength);
9        }
10   }
```

Figure 8.13 Queue

class provides a function called *rollback*. This function is called by an optimistic entity, whenever it receives a straggler message. The default actions of this function is to restore the entity using a checkpointed state with a timestamp immediately preceding the timestamp on the straggler message and begin recomputation. However, a derived entity can redefine this method such that if the straggler message is a *probe*, then the message is processed in the state that is saved at or immediately prior to the timestamp on the straggler. The subsequent events that have already been processed by the entity do not need to be canceled.

Commutative messages. It is sometimes possible to process straggler messages that modify the state of the recipient process without initiating a rollback; such a message is referred to as a *commutative* message. As an example, consider the following two sequences that

```
1    entity optimizedOptimisticQueue
2        inherit queue {
3        quelength(ename id) {
4            invoke id with result(qlength);
5        }
```

Figure 8.14 Derived entity

are input to a FIFO server: $r_1 = (5,10,LP_1),(18,7,LP_2),(30,8,LP_1)$ and $r_2 = (5,10,LP_1),(30,8,LP_1),(18,7,LP_2)$ where the message parameters respectively represent the message timestamp, desired service duration and the requesting LP. The final state of the server and the output message sequences to each customer are the same, regardless of which sequence of input messages is actually processed by the server; the message $(18,7,LP_2)$ is said to be commutative, and the permuted sequence r_2 is said to be *compatible* with the correct sequence.

Once again, it should be possible to redefine the *rollback* function in the derived entity, such that it is possible to detect commutative messages and process them *without* initiating a rollback.

The following example illustrates the use of the **warp** optimization available in the optimistic run time system. Upon receiving a job message, the entity simulates its service by executing an appropriate hold statement and then updating its job count it forwards the message on. The server object is defined below:

Further examination of the server object reveals that if a job message is received while the object is idle and that job can be serviced during this idle time then no roll back is necessary. All that is needed is to derive a new object that only adds a **warp** method. The code follows:

The new *rollback* method will determine if the message that was received during a time in which the object was idle. If so then the method can be called with the state saved at that point. After the method is processed then additional rollback is not necessary the object can continue with the next message received. If the object was not idle then we must call the *optimistic :: rollback* method which will rollback all computation and message invocations to the time of this message. By allowing the *rollback* method of the *optimistic* we are able to greatly reduce the recomputation needed in this case.

8.8. IMPLEMENTATION

MOOSE implementations are currently available for both sequential and parallel architectures. The sequential implementations may be executed on any hardware platforms that provide a C compiler. The parallel implementations may be executed on architectures supporting the Cosmic Environment[19] or SUN IPC. As with other object-oriented systems, the MOOSE compiler does not generate assembly code. Instead,

```
1    entity server {
2      int nojobs; ename creator;
3      int idle; ename rid;
4      local processjob(int id, int stim) {
5         idle = false;
6         hold(stim);
7         invoke rid with
8            job(id, stim); idle = true;
9         nojobs = nojobs + 1;
9      }
10     idmsg(ename id) {
11        rid = id;
12     }
13     job(int id, int stim) {
14        processjob(id, stim);
15     }
16     create(ename c) {
17        creator = c;
18     }
19   }
```

Figure 8.15 Server

a simple translator (or preprocessor) has been defined that translates a
MOOSE program into Maisie object code, which is subsequently compiled by the Maisie compiler for execution with a sequential, parallel
conservative or parallel optimistic algorithm. This section gives a brief
description of the MOOSE translator.

The translator is divided into two phases: the parsing phase which
translates each MOOSE object into an internal representation and the
code generation phase which emits a Maisie entity for each object. In
the parsing phase, each MOOSE object is represented internally as
a record structure where the local data structures and methods are
separated in distinct fields. The information about each method is
subdivided into three distinct components: the *message-type(s)*, the
corresponding *guard(s)* for each message-type and finally the code to
be executed which is stored as a *function* pointer. For derived objects,
the translator first copies the structure corresponding to the base object

and adds fields that correspond to new methods (if any) that are defined by the derived object. If any attributes of the inherited object are *redefined* in the derived entity, the new definitions are incorporated simply by copying over the old definition in the structure. By separating the definition of the guard, message, and function from the method we can allow the derived entity to *selectively* override any part of the method definition.

The code generation phase simply interprets the information in each structure to define a unique Maisie entity for each MOOSE object. The primary distinction between a Maisie entity and a MOOSE object is in the definition of language constructs to remove a message from the local message buffer. A Maisie entity uses a *wait* statement for this purpose. A wait statement consists of one or more *resume* statements. Each resume statement contains a resume condition and a compound statement, where the former is similar to the guard defined for a method, and the latter corresponds to the function body of the method. An object is essentially translated into an entity by defining a wait statement that is executed continuously within a single loop. Each method in the object is translated into an equivalent resume statement within the single wait statement. The translation process is illustrated by a simple example that translates the MOOSE object in Figure 8.17 into the equivalent Maisie entity of Figure 8.18.

The Maisie code is then compiled into C by the Maisie compiler. Interested readers are referred to the corresponding technical report[17] for further discussion of the Maisie implementations.

```
1    entity optimizedOptimisticServer
2    inherit server {
3    rollback() {
4        ...
5    }}
```

Figure 8.16 Parallel server: Derived entity

```
entity manager {
  int units;

  req(ename id) st(units > 0) {
    units − −;
    invoke id with done;
  }

  free() {
    units + +;
  }
}
```

Figure 8.17 The MOOSE Entity

8.9. CONCLUSION

This chapter described a parallel object-oriented simulation language called MOOSE. The basic goal of this work was to identify constructs that allow a programmer to use the principles of object-oriented design in developing simulation models that are extensible and whose performance can be optimized with specific simulation algorithms. MOOSE was designed by adding the constructs described here to an existing simulation language called Maisie.

Unlike previous languages which only provided partial support for inheriting the enabling conditions for an object's methods, MOOSE cleanly separates the method from its guard to allow each to be redefined separately. Each method defined in a MOOSE object may contain a guard that can specify dynamic enabling conditions which determine when the corresponding method may be executed. This synchronization information may be selectively inherited by a derived object. By allowing all attributes of an object, including its synchronization to be inherited, the MOOSE constructs provide full extensibility for its objects. In addition, MOOSE is the first simulation environment which uses inheritance to derive multiple versions of an object, where each version is optimized for execution with a specific simulation algorithm.

A simple translator has been implemented to translate MOOSE objects into Maisie entities. The resulting program may then be exe-

```
entity manager{} {
  int units;
  message req{ename id;};
  message free;

  for(;;) {
    wait until mtype(req) st(units > 0) {
      units − −;
      invoke msg.req.id with done;
    }
    or mtype(free) {
      units + +;
    }
  }
}
```

Figure 8.18 The Corresponding Maisie Entity

cuted using existing run-time systems developed for Maisie. This allows MOOSE simulations to be executed with a sequential algorithm, a parallel conservative algorithm as well as a parallel optimistic algorithm. The paper developed a simulation model of a stochastic benchmark in MOOSE and presented experimental results on the speedups that were obtained. The example was also used to illustrate the derivation of refined objects that were optimized for execution with specific simulation algorithms.

ACKNOWLEDGMENTS. The authors wish to thank Vikas Jha and Wen-toh Liao for their help in running the CQNF simulation. Gratitude is also due to the referees and Dr. Dennis Mok whose comments were useful in improving the text. We are also grateful to Professor Chuck Seitz for providing access to the Symult at Caltech, which was used for the measurements reported in this chapter. This research was partially supported under NSF PYI award ASC 9157610 and the ARPA/CSTO contract F-30602-94-C-0273, "Scalable Systems Software Measurement and Evaluation."

REFERENCES

1. RICHARD FUJIMOTO, "Parallel discrete event simulation," Communications of the ACM, October 1990, 33(10):30–53.

2. JAYADEV MISRA, "Distributed discrete-event simulation," Computing Surveys, March 1986, 18(1):39–65.

3. K. M. CHANDY and R. SHERMAN, "Space-time and simulation," In B. Unger and F. Fujimoto, editors, 1989 Simulation Multiconference: Distributed Simulation, Miami, Florida, 1989, pages 557–561.

4. M. L. NELSON, "Concurrency and object-oriented programming," ACM SIGPLAN Notices, October 1991, 26(10):63–72.

5. PIERRE AMERICA and FRANK VAN DER LINDEN, "A parallel object-oriented language with inheritance and subtyping," In ECOOP/OOPSLA 1990 Proceedings Conference on Object-Oriented Programming: Systems, Languages, and Applications European Conference on Object-Oriented Programming, October 1990, pages 161–168.

6. D. G. KAFURA and K. H. LEE, "Inheritance in actor based concurrent object-oriented languages," The Computer Journal, 1989, 32(4):297–304.

7. CHRIS TOMLINSON and VINEET SINGH, "Inheritance and synchronization with enabled-sets," In Norman Meyrowitz, editor, OOPSLA '89 Object-Oriented Programming: Systems, Languages and Applications, October 1989, pages 103–112.

8. EDWARD C. RUSSELL, "Simscript II.5 and modsim II: A brief introduction," In Gordon M. Clark Barry L. Nelson, W. David Kelton, editor, Proceedings of the 1991 Winter Simulation Conference, 1991, pages 62–66.

9. GREG LOMOW and DIRK BAEZNER, "A tutorial introduction to object-oriented simulation and sim++," In Gordon M. Clark Barry L. Nelson, W. David Kelton, editor, Proceedings of the 1991 Winter Simulation Conference, 1991, pages 157–163.

10. BJARNE STROUSTRUP, The C++ Programming Language. Addison-Wesley Publishing Company, July 1987.

11. BERTRAND MEYER, Object-oriented Software Construction. Prentice Hall, 1988.

12. R. L. BAGRODIA, K. M. CHANDY, and W. LIAO, "An experimental study on the performance of the *Space-Time* simulation algorithm," In Proceedings of 6th Workshop on Parallel and Distributed Simulation, January 1992.

13. RANDAL E. BRYANT, "Simulation of packet communication architecture computer systems," Technical Report MIT-LCS-TR-188, MIT, 1977.

14. DAVID R. JEFFERSON, "Virtual time," ACM Transactions on Programming Languages and Systems, July 1985, 7(3):404–425.

15. R. L. BAGRODIA, K. M. CHANDY, and W. LIAO, "A unifying framework for distributed simulations," ACM Transactions on Modeling and Computer Simulation, October 1991, pages 348–385.

16. YI-BING LIN and EDWARD D. LAZOWSKA, "Exploiting lookahead in parallel simulation," IEEE Transactions on Parallel and Distributed Systems, October 1990, 1(4):457–469.

17. RAJIVE BAGRODIA and WEN-TOH LIAO, "A language for iterative design of efficient simulations," Technical report No. UCLA-CSD-920044, Computer Science Department, UCLA, Los Angeles, CA 90024, October 1992.

18. R. L. BAGRODIA and VIKAS JHA, "Transparent implementation of conservative algorithms in parallel simulation languages," In   Proceedings of 1993 WCS Multiconference on Distributed Simulation, January 1993.

19. C.L. SEITZ, J. SEIZOVIC, and WEN-KING SU, The C Programmer's Abbreviated Guide to Multicomputer Programming, Technical Report Caltech-CS-TR-88-1, Dept. of Computer Sciences, California Institute of Technology, Los Angeles, January 1988.

FORMAL SYNTAX

The syntax of MOOSE statements is given in BNF using the following symbols and conventions:

terminals	:	terminal symbols are given in **boldface** style
nonterminals	:	nonterminal symbols are in *italic* style
ident	:	an identifier
[]	:	symbols occurring within brackets are optional
[]...	:	symbols can occur 0 or more times
|	:	alternative

entity-def   ::= **entity entity-type** *inheritance* { [*member*]... }

inheritance   ::= **inherit entity-type**

member   ::= *message* | *data-declaration* | *constructor* | *destructor* | *method*

message   ::= **message message-type** ([*parameter*]...)

data-declaration   ::= **ename ident**
          | a Maisie data declaration

constructor   ::= **create** ([*parameter*]...) [*statement*]...

destructor   ::= **destroy** ([*parameter*]...) [*statement*]...

method   ::= [**method-name**] [*message-def*]... [**st** (*Boolean-expr*)] [*statement*]

message-def   ::= **message-type**([*parameter*]...)

parameter   ::= **ename entity-ident**
          | simple Maisie parameter

statement   ::= **entity-ident = new entity-type**
          | **destroy entity-ident**
          | **invoke entity-ident with message-type** ([*parameter*]...)
          | a Maisie statement
Boolean-expression ::= a Maisie Boolean expression

FORMAL SYNTAX

The syntax of MOOSE statements is given in BNF using the following symbols and conventions.

terminals	terminal symbols are given in boldface style
nonterminals	nonterminal symbols are in italic style
ident	an identifier
{}	symbols occurring within brackets are optional
[]	symbols can occur 0 or more times
...	alternative

entity-def	::= entity entity-type * inheritance { [member] ...}
inheritance	::= inherit entity-type
member	::= message	data-declaration	constructor	destructor	method
message	::= message message-type (parameters)
core-declaration	::= create class
	::= Maisie data declaration
constructor	::= create ([parameters ...]) [statement]
destructor	::= destroy ([parameters ...]) [statement]
method	::= [method-name] (statement ...) [... [Boolean-expr)] statement]
message-def	::= message-type [parameters ...]
parameter	::= volume entity-ident
	::= simple Maisie parameter
statement	::= entity-ident = new entity-type
	::= destroy entity-ident
	::= invoke entity-ident with message-type [parameters ...]
	::= Maisie statement
Boolean-expression	::= Maisie Boolean expression

CHAPTER 9
STOCHASTIC PETRI NETS: INTRODUCTION AND APPLICATIONS TO THE MODELING OF COMPUTER AND COMMUNICATION SYSTEMS

Gianfranco Ciardo

9.1. INTRODUCTION

Complex discrete-state systems can be modeled directly by specifying a state-to-state transition diagram, but doing so is seldom feasible in practice because of the enormous size of the resulting description. High-level formalisms are normally employed instead. The compactness of the description expressed in one of these formalisms is in itself extremely desirable, since it saves specification time, makes visual inspection easier, and reduces the overall likelihood of modeling errors. The SPN formalism has the additional advantage of allowing specialized analysis techniques to be employed directly "at the net level", thus resulting in efficient (in term of memory and time complexity) algorithms.

SPN's have gained wide acceptance for the modeling of computer and communication systems because of their extreme flexibility combined with a small set of basic rules defining their semantics. Starting from the simple definition of (untimed) Petri nets, various classes of SPN's can be obtained, with varying power and complexity, according to the timing and stochastic characteristics of the system being modeled. More powerful classes of SPN's require increasingly difficult and expensive algorithms for their solution. On the one end, SPN's where all activities have an exponentially-distributed duration are the simplest

203

to define and solve, but can model exactly only systems whose behavior can be represented as a continuous-time Markov chain (CTMC). At the opposite extreme, SPN's with arbitrary distributions can be defined, which can model virtually any discrete-state system, but these require to specify many more behavioral characteristics which can be instead ignored in the previous case; discrete-event simulation is the only practical solution approach for these models.

The modeling power and the analysis techniques used for a given class of SPN's are determined by the stochastic process underlying the SPN, since it is this process that is ultimately analyzed. Starting from the basic Petri net definition, in Section 9.2, we consider SPN's having an underlying CTMC or discrete-time Markov chain (DTMC), in Sections 9.3 and 9.4, respectively. Then, we move to more sophisticated classes of SPN's having underlying processes which are either semi-Markov or Markov-regenerative, in Section 9.5. Finally, Section 9.6 surveys more specialized areas of recent SPN research. Running examples are used throughout the presentation, and appropriate bibliographic citations are given.

9.2. PETRI NETS

In this section, we define the basic Petri net model and some structural extensions commonly employed. Then, we describe the applications which will be used throughout the presentation. Petri nets were introduced by C. A. Petri in 1962[46]. A Petri net is a bipartite directed weighted graph with two sets of nodes: places and transitions. Arcs connect places to transitions or transitions to places. The weight of an arc is called its cardinality. Graphically, places are represented by circles, while transitions are represented by rectangles or bars. An arc from place p to transition t (from t to p) is an *input* (*output*) arc of t. We also say that p is an *input* (*output*) place of t.

Places contain tokens, drawn as black circles. If P is the set of places, a *marking* is a vector, $\mu \in \mathbb{N}^{|P|}$, describing the number of tokens in each place. μ_p is the number of tokens in place p in marking μ.

Formally, a Petri net is a 5-tuple $\left(P, T, D^-, D^+, \mu^{[0]}\right)$ where:

- $P = \{p_1, ..., p_{|P|}\}$ is a finite set of places.

- $T = \{t_1, ..., t_{|T|}\}$ is a finite set of transitions $(P \cap T = \emptyset)$.

- $D^- \in \mathbb{N}^{|P \times T|}$ and $D^+ \in \mathbb{N}^{|P \times T|}$ describe the input and output arcs. $D_{p,t}^-$ is the cardinality of the input arc from p to t and $D_{p,t}^+$ is the cardinality of the output arc from t to p.

- $\mu^{[0]} \in \mathbb{N}^{|P|}$ is the initial marking.

A transition t is *enabled* in marking μ iff each of its input places contains at least as many tokens as the cardinality of the corresponding arc, $\mu \geq D_{p,\bullet}^-$. An enabled transition may then *fire* by removing tokens from its input places and depositing tokens in its output places, again according to the cardinality of the corresponding arcs. The firing of t in μ changes the marking to μ', we write $\mu \xrightarrow{t} \mu'$, where $\mu' = \mu + D_{p,\bullet}^+ - D_{p,\bullet}^-$. Let $\mathcal{E}(\mu) \subseteq T$ be the set of transitions enabled in marking μ.

We extend the notation to *firing sequences*: if $s = (t_1, t_2, \ldots t_n) \in T^*$ is a sequence of transitions, $\mu \xrightarrow{s} \mu'$ indicates that s can be fired starting in μ, that is, $\exists \mu_1, \mu_2, \ldots \mu_{|s|-1} \in \mathbb{N}^{|P|}$ such that $\mu \xrightarrow{t_1} \mu_1$, $\mu_1 \xrightarrow{t_2} \mu_2$, and so on, until μ' is reached by firing t_n. A marking μ is *reachable* iff there is a firing sequence s such that $\mu^{[0]} \xrightarrow{s} \mu$. The *reachability set* is the set of all reachable markings: $\mathcal{R} = \{\mu \mid \exists s \in T^*, \mu^{[0]} \xrightarrow{s} \mu\}$. The *reachability graph* $(\mathcal{R}, \mathcal{A})$ is the graph having the reachable markings as nodes, and an arc from μ to μ', annotated with t, if $\mu \xrightarrow{t} \mu'$.

Firing is an atomic event. If two transitions are enabled in a Petri net, they cannot be fired "at the same time", a choice is made concerning which one to fire first, the other can only fire after that, if it is still enabled. This is not a restriction; when transitions are simultaneously enabled, the modeled activities are proceeding in parallel. The first transition to fire simply represents the first activity to complete. Petri net semantics do not state which of two simultaneously enabled transitions must fire first.

Petri nets can be studied using two main analytical tools: structural analysis and reachability (graph) analysis. The former is concerned with obtaining behavioral information from the *incidence matrix*, defined as $D = D^+ - D^-$, so that $D_{\bullet,t}$ is the net marking change when t fires. The fundamental property of D is that $\mu \xrightarrow{s} \mu'$ implies that $\mu' = \mu + Df(s)$ where $f(s) \in \mathbb{N}^{|T|}$ is the *firing count vector* for the sequence s, that is, the number of times each transition $t \in T$ appears in s. A necessary condition for μ to be reachable is that $\mu' - \mu^{[0]} = Dy$ has a solution $y \in \mathbb{N}^{|T|}$. Unfortunately, this is not a sufficient condition, so structural analysis has limitations. Of special interest are the *t-semiflows*[25], nonzero solutions $y \in \mathbb{N}^{|T|}$ to the equation $Dy = 0$. Any firing sequence s having y as a firing count is guaranteed to leave the marking unchanged: if s is firable in μ, then $\mu \xrightarrow{s} \mu$. Since any linear combination of solutions to $Dy = 0$ is itself a solution, only the *minimal* t-semiflows are normally computed.

P-semiflows[25], the nonzero solutions $x \in \mathbb{N}^{|P|}$ to $xD = 0$, also have important properties. In any reachable marking, the sum of the number of tokens in each place weighted according to x is a constant:

$$\forall \mu \in \mathcal{R}, \sum_{p \in P} x_p \mu_p = \sum_{p \in P} x_p \mu_p^{[0]}.$$

If there is a p-semiflow with $x_p > 0$, we say that p is covered (by a p-semiflow), and this ensures that the number of tokens in p is *bounded* in any reachable marking. Petri nets where all places are covered are of particular interest, since they have a finite reachability set.

Reachability analysis uses instead an exhaustive approach to study the Petri net. While some limited analysis is possible even when $\mathcal{R}$ is infinite, this technique is mostly applied assuming finite reachability sets. Even then, the size of $\mathcal{R}$ might preclude its generation and storage. The algorithm in Fig. 9.1 illustrates the main steps to build $\mathcal{R}$, assuming it is finite; basically, it performs a breadth-first search of the graph implicitly defined by the Petri net. Such an analysis, when feasible, provides complete information to answer questions regarding the Petri net. Structural analysis can instead provide only partial information based on the incidence matrix, but it cannot take into account the initial marking, nor the more powerful extensions discussed in the next section. Indeed, the incidence matrix cannot even capture the effect of a *control place*, a place p which is both input and output for a transition t, since the entries $D_{p,t}^-$ and $D_{p,t}^+$ cancel each other in this case.

For more detailed information, we refer to the rich existing literature on Petri nets, special subclasses of Petri nets, and their properties[43,45,49].

9.2.1. Petri net extensions

Several structural extensions of basic Petri nets have been proposed to extend their flexibility and modeling power. It is well known that Petri nets lack *zero-testing*, the ability to enable a transition only when a given (unbounded) place is empty. *Inhibitor arcs, transition priorities*, and *guards* (also known as enabling or inhibiting functions) were introduced to overcome this limitation. Unfortunately, the resulting (extended) Petri net formalism becomes Turing-equivalent, thus its analysis is severely hampered.

Inhibitor arcs connect a place to a transition and are drawn with a small circle instead of an arrowhead. An inhibitor arc from place p to

$\mathcal{R} \leftarrow \{\mu^{[0]}\}; \quad \mathcal{A} \leftarrow \emptyset; \quad \mathcal{U} \leftarrow \{\mu^{[0]}\}; \qquad$ /* unexplored markings, $\mathcal{U} \subseteq \mathcal{R}$ */
while $\mathcal{U} \neq \emptyset$ do
    choose a marking μ from $\mathcal{U}$;
    $\mathcal{U} \leftarrow \mathcal{U} \setminus \{\mu\}$;
    for each $t \in T$ do
        if t is enabled in μ then
            $\mu' \leftarrow \mu + D^{+}_{\bullet,t}(\mu) - D^{-}_{\bullet,t}(\mu); \quad \mathcal{A} \leftarrow \mathcal{A} \cup \{(\mu \xrightarrow{t} \mu')\}$;
            if $\mu' \notin \mathcal{R}$ then
                $\mathcal{U} \leftarrow \mathcal{U} \cup \{\mu'\}; \quad \mathcal{R} \leftarrow \mathcal{R} \cup \{\mu'\}$;

Figure 9.1: Algorithm to generate the reachability graph $(\mathcal{R}, \mathcal{A})$.

transition t with cardinality $c > 0$ disables t in any marking where p contains at least c tokens. Priorities have been defined as a nonnegative integer assignment to each transition: a transition may be enabled only if no higher-priority transition is enabled. We prefer to define priorities with an acyclic relation "$\succ$" over T: if $t \succ u$, u is disabled whenever t is enabled. This relation can be represented graphically, by drawing an arc from t to u (we use a small circle on the destination as for inhibitor arcs, to stress the disabling effect of the priority). The main characteristic of this definition of priority is that transitivity is not required: $t \succ u$ and $u \succ v$ do not imply $t \succ v$, while this is not the case with integer priorities.

The priority relation is static, does not depend on the current marking. If a transition needs to be disabled according to *marking-dependent* conditions, we can associate guards g, boolean predicates on the marking, to the transitions. A transition t is enabled in marking μ only if its guard $g_t(\mu)$ is satisfied. The flexibility of guards is offset by their nongraphical nature, which makes the model harder to understand.

A final extension, marking-dependent arc cardinalities, can simply increase flexibility, or achieve Turing-equivalence, depending on specific restrictions. In the most general case, we define D^-, D^+, and D° (for inhibitor arcs) to be matrices of marking-dependent functions, instead of natural numbers. A common use of this feature is to define the cardinality of the input arc from p to t as the number of tokens in p itself, $D^-_{p,t}(\mu) = \mu_p$, so that the firing of t empties p. Inhibitor arcs can be modeled by marking-dependent arc cardinalities: the effect of $D^-_{p,t} = c_1$ and $D^\circ_{p,t} = c_2$ is achieved by defining $D^-_{p,t}(\mu) = $ "$\mu_p + 1$ if $\mu_p \geq c_2$, c_1 otherwise". This extension preserves the ability to perform structural analysis if the marking-dependency is limited to nonhomogeneous linear combinations of the marking, that is, if

$$D_{p,t}^-(\mu) = \alpha_{p,t}^- + \sum_{q \in P} \beta_{p,t,q}^- \mu_q \quad \text{and} \quad D_{p,t}^+(\mu) = \alpha_{p,t}^+ + \sum_{q \in P} \beta_{p,t,q}^+ \mu_q$$

for some $\alpha \in \mathbb{N}^{|P| \times |T|}$ and $\beta \in \mathbb{N}^{|P| \times |T| \times |P|}$ (self-modifying nets[54]). It has been shown that an equivalent ordinary Petri net can be derived, from which the p-semiflows can be computed[14].

Other extensions to Petri nets, which we do not include in our definition, are debit arcs and anti-tokens[53], colored tokens[37], and predicate-transition nets[49] (the last two fall in the category of *high-level nets*).

In the special case of Petri nets with finite reachability sets, all the Turing extensions merely increase convenience and do not extend the class of systems that may be represented beyond that of finite state machines. This is often the case of interest to us.

9.2.2. Application: a transmission protocol

We consider a simple transmission protocol where a local processor continuously generates messages, sends them to a remote processor, and waits for acknowledgments (a related model has been discussed elsewhere[18]). If the acknowledgment does not reach the local processor within a certain amount of time after the original transmission started, a timeout occurs. Fig. 9.2 shows a SPN for this system. Normally, the local processor is running (a token is in place *local*) and no outstanding message or acknowledgment exists. Periodically, the local processor generates a message for transmission (transition *gen* fires and deposits a token in in place *msg*). When the message leaves the local processor (transition $xmit_M$ fires), a timeout counter is also started (a token is placed in place *wait*). Eventually, the message (token in place $transit_M$) reaches the remote processor (transition rcv_M fires), which immediately starts sending an acknowledgment (transition $xmit_A$). The acknowledgment (token in place $transit_A$) is eventually received by the local processor (transition rcv_A fires) and the timeout counter is reset (the token in *wait* is removed).

If, on the other hand, the message or its acknowledgment encounter delays, a timeout can occur (transition *tout* fires), the transmission is deemed unsuccessful, and the message is immediately retransmitted (a token is put back in place *msg*). In practice, this implies the ability to ignore late acknowledgments received by the local processor and duplicate messages received by the remote processor, perhaps through the use of sequencing numbers. In our Petri net model, this is

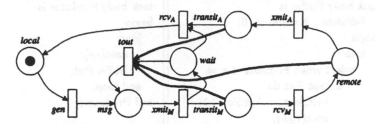

Figure 9.2: The Petri net model for the transmission protocol.

Table 9.1: $\mathcal{R}$ and minimal semiflows for the transmission protocol SPN.

Marking	local	remote	msg	transit$_A$	wait	transit$_M$
0	1	0	0	0	0	0
1	0	0	1	0	0	0
2	0	0	0	0	1	1
3	0	1	0	0	1	0
4	0	0	0	1	1	0

P-semiflow	local	remote	msg	transit$_A$	wait	transit$_M$
p_1	1	0	1	0	1	0
p_2	1	1	1	1	0	1

T-semiflow	gen	xmit$_M$	rcv$_M$	xmit$_A$	rcv$_A$	tout$_a$	tout$_b$	tout$_c$
t_1	1	1	1	1	1	0	0	0
t_2	0	1	0	0	0	1	0	0
t_3	0	1	1	0	0	0	1	0
t_4	0	1	1	1	0	0	0	1

achieved through the use of variable-cardinality arcs. The input arcs from $transit_M$, $remote$, and $transit_A$ to $tout$, drawn with thick lines, have cardinalities equal to the number of tokens in the input place, that is, they are "flushing arcs". This ensures that the token residing in one of those three places is removed when a timeout occurs. Without those three input arcs, the reachability set would be infinite, since a token is added to place $transit_M$ each time the sequence $(xmit_M, tout)$ fires. A reachability analysis of this Petri net shows that $\mathcal{R}$ contains the five markings listed in Table 9.1. From a structural analysis, we can obtain the minimal semiflows shown in Table 9.1. The columns $tout_a$, $tout_b$, and $tout_c$ correspond to possible positions of the token, in $transit_M$, $remote$, and $transit_A$, respectively, when transition $tout$ fires.

```
task body Buffer is                    task body Producer is
   FullSlots : Natural := 0;           begin
begin                                     loop
   loop                                      exec(i_P);
      select when FullSlots < n =>           Buffer.Put;
         accept Put do                     end loop;
            FullSlots := FullSlots+1;    end Producer;
            exec(i_BP);
         end Put;
      or when FullSlots > 0 =>          task body Consumer is
         accept Get do                  begin
            exec(i_BC);                    loop
            FullSlots := FullSlots-1;         Buffer.Get;
         end Get;                            exec(i_C);
      end select;                         end loop;
   end loop;                            end Consumer;
end Buffer;
```

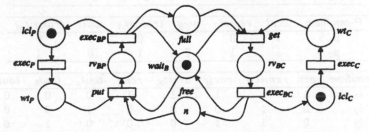

Figure 9.3: Ada code and Petri net model for a producer-consumer system.

9.2.3. Application: modeling Ada tasking systems

Consider an Ada tasking system where data items produced by a producer task are passed to a buffer task and eventually consumed by a consumer task, as shown in the top portion of Fig. 9.3 (a related model has been discussed elsewhere[21]). The buffer task stores the incoming items into an internal array with n positions, and it uses the integer variable FullSlots to keep track of the number of nonempty slots. The producer task cannot pass items to the buffer task when FullSlots is equal to n and the consumer task cannot retrieve items from the buffer task when FullSlots is equal to 0. Hence, n affects the size of the window between the number of produced and consumed items, but the production and consumption rates must be equal in steady-state.

The "rendezvous" is the Ada mechanism used for task synchronization. Whenever a producer task has an item ready to pass, it issues the "entry call" Buffer.Put to the buffer task. When the buffer task accepts this entry call, the rendezvous takes place, FullSlots is incremented, and the item is copied into the internal array. Similarly, a rendezvous with a consumer decrements FullSlots by one. The "exec(k)" statements signify the execution of k machine-level instructions. A Petri net for this system is shown at the bottom of Fig. 9.3.

Reachability analysis of this Petri net can be performed for each choice of n. For $n = 1, 2, 3$, we obtain 12, 20, and 28 markings respectively. To obtain a general expression for the size of the reachability set as a function of n, as, we can observe the following:

- When there is a token in wt_B, the number of tokens in $free$ can vary from zero to n, while, at the same time, the token initially in lcl_P can be either there or in wt_P, and the one in lcl_C can be either there or in wt_C, for a total of $4(n + 1)$ markings.

- When there is a token in rv_{BP}, the number of tokens in $free$ can vary from zero to $n - 1$, while the token initially in lcl_C can be either there or in wt_C, for a total of $2n$ markings.

- Analogously, we have $2n$ markings with a token in rv_{BC}.

Hence, the total number of markings as a function of n is $8n + 4$.

Table 9.2 shows the minimal semiflows for this SPN model. If we wanted to omit the restriction of a finite-size array, we could simply remove place $free$ from the model. This results in an infinite reachability set, as place $full$ becomes unbounded. Nevertheless, the same t-semiflow covering all the transitions exists, thus structural analysis is still able to show that the number of firings of any transition in the net must be asymptotically the same, for a stable system.

9.3. CONTINUOUS-TIME MARKOV SPN'S

Time can be modeled with Petri nets by attaching a *firing time* to the transitions, corresponding to the duration of the activity modeled. When these are described by a probability distribution, the resulting model is a SPN. If all the distributions are strictly continuous with support over $[0, \infty)$, every marking reachable in the untimed Petri net is also reachable in the SPN. This property greatly simplifies the logical

Table 9.2: The minimal semiflows for the producer-consumer PN.

P-semiflow	ld_P	wt_P	rv_{BP}	ld_C	wt_C	rv_{BC}	wt_B	$full$	$free$
p_1	1	1	1	0	0	0	0	0	0
p_2	0	0	0	1	1	1	0	0	0
p_3	0	0	1	0	0	1	1	0	0
p_4	0	0	0	0	0	0	0	1	1

T-semiflow	ex_P	put	ex_{BP}	ex_C	get	ex_{BC}
t_1	1	1	1	1	1	1

Table 9.3: The firing time distributions for the producer-consumer SPN.

ex_P	put	ex_{BP}	ex_C	get	ex_{BC}
Expo(1)	Expo(1000)	Expo(10)	Expo(1)	Expo(1000)	Expo(10)

analysis. If we can also assume that the distributions are memoryless, hence exponential, only the *firing rate* needs to be specified and the evolution of the SPN in a marking μ depends only on μ, not on how long the SPN has been in μ. SPN's of this type define an underlying stochastic process which is a CTMC[42,44]. This initial class of SPN's was extended by allowing zero firing times (which can be thought of as exponential distributions with infinite rate), resulting in the extremely successful GSPN formalism[3,4,5,9]. Another extension is the use of (continuous) phase-type distributions[1], which can themselves be expressed as the time to absorption in a finite CTMC. In both cases, however, the bijection between markings of the SPN and states of the CTMC is lost. If *immediate transitions* with zero firing time exist, any marking enabling them, said to be *vanishing*, does not have a corresponding state in the CTMC. On the other hand, a marking enabling a transition with a phase-type distribution corresponds to multiple states in the CTMC.

The advantages of immediate transitions certainly justify the added complexity. For example, consider the SPN in Fig. 9.3 when the transitions have firing times distributed as in Table 9.3. The resulting CTMC has $8n + 4$ states and $14n + 4$ nonzero entries. It is apparent, though, that transitions *put* and *get* have a much higher rate than the others. They correspond to activities which are almost instantaneous compared to the other transitions. This was a motivation for introducing immediate transitions. If the distributions of *put* and *get* are changed to Const(0), we obtain a GSPN with $5n + 3$ *tangible* markings (enabling only *timed transitions*) and $2n + 2$ vanishing markings. Its underlying

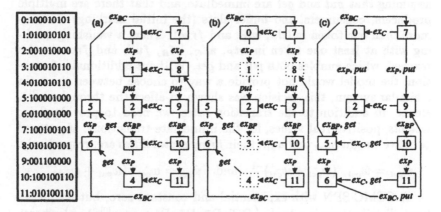

Figure 9.4: The state space for the producer-consumer system.

CTMC has $5n + 3$ states and $8n + 4$ nonzero entries. Using immediate transitions instead of timed transitions with a high rate results in:

- an underlying CTMC matrix with fewer states and, normally, fewer nonzeros as well (but there are pathological cases were the number of nonzeros can actually increase[20]), and,

- possibly, a less stiff linear system; if so, the numerical solution of the GSPN is then potentially more stable.

If we solve both the SPN and the GSPN with $n = 100$, we learn that the transitions have a throughput of 0.89356 and 0.89443, respectively (this is the rate at which data items are produced and consumed). The small difference, due to the different timing for *put* and *get*, is a small price to pay, since it is offset by a shorter solution time. This is true for most applications, where the growth of the state space is combinatorial in the number of tokens, and the selected use of immediate transitions can substantially reduce the order of growth. For illustration purposes, Fig. 9.4 shows (a) the reachability graph for the SPN assuming all exponentially-distributed transitions; (b) the reachability graph for the GSPN, where vanishing markings are drawn with dotted lines; and (c) the *reduced reachability graph*, where the vanishing markings have been removed (this is the underlying CTMC if the appropriate rates are substituted to the transition names).

The introduction of immediate transitions requires additional specifications to fully describe the model behavior. Consider again Fig. 9.3, assuming that *put* and *get* are immediate, and that there are multiple producers, consumers, and buffer slots (the initial markings contains more than one token in lcl_P, wt_C, and $free$). Then, a vanishing marking with at least one token in wt_P, wt_C, wt_B, $full$, and $free$ can be reached, which enables both *put* and *get*. Without additional information, the model would not provide a way to choose between them (in a similar system, this decision was shown to affect the throughput[21]). Hence, in addition to the transition rates, we need to provide *firing weights*, positive real values, for the immediate transitions. If *put* and *get* have weight w_{put} and w_{get}, their *firing probabilities* are, respectively,

$$\hat{w}_{put} = w_{put} \cdot (w_{put} + w_{get})^{-1} \quad \text{and} \quad \hat{w}_{get} = w_{get} \cdot (w_{put} + w_{get})^{-1}.$$

A CTMC-SPN with exponential and constant-zero, but no phase-type, distributions is a tuple $\left(P, T, D^-, D^+, D^\circ, \succ, g, \mu^{[0]}, \lambda, w\right)$ where:

- $\left(P, T, D^-, D^+, D^\circ, \succ, g, \mu^{[0]}\right)$ define an extended Petri net.

- $\forall \mu \in \mathbb{N}^{|P|}, \forall t \in \mathcal{E}(\mu), \lambda_t(\mu) \in \mathbb{R}^+ \cup \{\infty\}$ is the rate of the exponential distribution for the firing time of transition t. If $\lambda_t(\mu) = \infty$, the firing time of t in μ is zero. In the original definition of GSPN's, transitions were *a priori* classified as timed or immediate; we base this choice on the current marking instead. Let $\mathcal{T}$ and $\mathcal{V}$ be the tangible and vanishing portion of the reachability set, respectively: $\mathcal{T} \cup \mathcal{V} = \mathcal{R}$. We assume that, whenever an immediate transition is enabled, no timed transition is enabled.

- $\forall \mu \in \mathbb{N}^{|P|}, \forall t \in \mathcal{E}(\mu), w_t(\mu) > 0$ is the firing weight assigned to enabled transition t. This weight is relevant only when $\lambda_t(\mu) = \infty$. The firing probability of immediate transition $t \in \mathcal{E}(\mu)$ is

$$\hat{w}_t(\mu) = w_t(\mu) \cdot \left(\sum_{u \in \mathcal{E}(\mu)} w_u(\mu)\right)^{-1}.$$

9.3.1. Solution algorithms

This section describes how to solve a CTMC-SPN, assuming that no phase-type distributions are used. The implications of phase-type distributions in the continuous and discrete case are similar, hence we discuss them only for the latter, in Section 9.4.

The algorithm of Fig. 9.1 can be augmented to generate $\mathcal{T}$, $\mathcal{V}$, and

$$A = \begin{bmatrix} A_{\mathcal{VV}} & A_{\mathcal{VT}} \\ A_{\mathcal{TV}} & A_{\mathcal{TT}} \end{bmatrix},$$

where A_{VV}, A_{VT}, A_{TV}, and A_{TT} are the transition probabilities from each vanishing marking to each vanishing and tangible marking, and the transition rates from each tangible marking to each vanishing and tangible marking, respectively. These are obtained as the sum of all the probabilities (or rates) for the enabled transitions causing the corresponding change of marking. The transition rate matrix R of the underlying CTMC is obtained as[4] $R = A_{TT} + A_{TV}(I - A_{VV})^{-1}A_{VT}$, and the infinitesimal generator Q is obtained from R by setting its diagonal entries to the negative of the sum of the off-diagonal entries of the corresponding row. The initial probability vector $\pi(0)$ is given by $\pi(0) = e_T + e_V(I - A_{VV})^{-1}A_{VT}$, where the vector $e = [e_V|e_T]$ contains all zero entries except a one in the position corresponding to the initial marking, which can be either vanishing or tangible.

If the CTMC has a single recurrent class, we can compute the steady-state instantaneous probabilities, π^*, by solving the equation $\pi^*Q = 0$ subject to $\sum_{i \in T} \pi_i^* = 1$. Note that π^* is independent of $\pi(0)$ in this case. If there are transient states, the expected cumulative sojourn times spent in them, σ^*, can be of interest. These are computed by solving the equation $\sigma^*\overline{Q} = -\overline{\pi}(0)$, where $\overline{Q}$ and $\overline{\pi}(0)$ are the restrictions of Q and $\pi(0)$ to the transient markings. Regardless of the CTMC structure, transient analysis, the study of the probability vector $\pi(\theta)$ or accumulated sojourn times $\sigma(\theta)$ up to time θ, is always meaningful. $\pi(\theta)$ is obtained by solving the Kolmogorov ordinary differential equation $\dot{\pi}(\theta) = \pi(\theta)Q$ with initial condition $\pi(0)$, where $\dot{\pi}(\theta)$ is the derivative of $\pi(\theta)$ with respect to θ. $\sigma(\theta)$ is obtained as the solution of the analogous Kolmogorov differential equation, $\dot{\sigma}(\theta) = \sigma(\theta)Q + \pi(0)$ with initial condition $\sigma(0) = 0$, obtained by integrating the previous equation with respect to θ. Several numerical methods exist. Fast iterative methods exploiting the sparsity of the matrices involved are used for steady-state solution, such as successive over-relaxation[52] (SOR) or multilevel methods[34]. Uniformization, or Jensen's method, is the method of choice for transient analysis[29].

At times, the action of *entering* a state, rather than *being* in it, is the focus of the analysis. Then, the frequency at which each tangible recurrent marking i is entered in the ergodic case is given by $\phi_i^* = \pi_i^*(-Q_{i,i})$, while the expected number of times a tangible transient marking i is entered is given by $n_i^* = \sigma_i^*(-Q_{i,i})$. The corresponding quantities for the vanishing markings are computed by solving for ϕ_V in

$$\phi_V^*[I - A_{VV}] = \phi_T^* A_{TV} \quad \text{and for } n_V \text{ in} \quad n_V^*[I - \overline{A}_{VV}] = n_T^* \overline{A}_{TV},$$

where the subscript have the obvious meaning. Analogous relationships exist for the transient counterparts $\phi(\theta)$ and $n(\theta)$.

The measures of interest are defined by a *reward structure*[35] (ρ, r):

- $\rho : \mathbb{N}^{|P|} \to \mathbb{R}$ describes the reward rates of the markings: a sojourn of duration x in marking μ accumulates a reward $\rho(\mu)x$.

- $r : T \times \mathbb{N}^{|P|} \to \mathbb{R}$ describes the reward impulses gained by firing transitions: $r_t(\mu)$ is the reward gained by firing transition t in marking μ.

Then, the *accumulated reward* process up to time θ is

$$Y(\theta) = \int_0^\theta \rho(\mu(u))du + \sum_{n=1}^{\max\{n:\theta^{[n]}\leq\theta\}} r_{t^{[n]}}(\mu^{[n]}),$$

where $\mu(u)$ is the (tangible) marking at time u and $t^{[n]}$, $\theta^{[n]}$, and $\mu^{[n]}$ are the n-th transition to fire, the time at which it fires, and the marking where the firing occurs, respectively. If all impulse rewards are zero, the *instantaneous reward rate* at time θ, $y(\theta) = Y(\theta)$ can also be defined.

Using an appropriate choice for (ρ, r), we can describe many measures of interest, from simple ones such as the number of tokens in a place or the number of firings of a transition, to complex ones involving the entire marking and the enabled transitions. The expected value of this stochastic processes in steady-state, or at a given time, is obtained from π^*, ϕ^*, σ^*, n^*, or their transient counterparts.

For example, the expected number of tokens in place p in steady state is given by $\sum_{\mu \in \mathcal{T}} \pi_\mu^* \rho(\mu)$, if we define $\rho(\mu) = \mu_p$. The mean-time-to-absorption (MTTA) in an SPN where all recurrent markings are absorbing is given by $\sum_{\mu \in \mathcal{T}} \sigma_\mu^* \rho(\mu)$, if we define $\rho(\mu) = 1$ for transient markings, 0 otherwise.

It should be noted that the above description corresponds to eliminating the vanishing markings "after the reachability graph generation". In practical implementations, alternative approaches are possible. If reducing the memory requirements is critical, as it often is, vanishing markings could be eliminated "on the fly". This corresponds to following a *vanishing path* in the reachability graph: if $\mu \xrightarrow{t} \mu_1 \xrightarrow{t_1} \cdots \xrightarrow{t_k} \mu'$, and all intermediate markings except μ and μ' are vanishing, we can simply add $\lambda_t(\mu)\hat{w}_{t_1}(\mu_1) \cdots \hat{w}_{t_k}(\mu_k)$ to the entry of Q corresponding to the rate of going from μ to μ'. This allows to store in its entirety only $\mathcal{T}$, and it is usually the best approach, although each vanishing marking is examined once for each different vanishing path containing it. While this can add an exponential complexity to the analysis in a worst-case scenario, it is seldom a problem in practice. However, it is more difficult to deal with cycles of vanishing markings, or *vanishing loops*, which sometimes occur in real models[13,31].

A diametrically opposite approach is possible, which does not eliminate the vanishing markings at all, and it is thus called *preservation*. It is based on the observation that, if we embed the GSPN at the time of transition firings, we obtain a DTMC with state space $\mathcal{R}$. $\Pi_{i,j}$ describes the probability of going from marking i to marking j in one transition firing, regardless of whether the transition is immediate or timed. Π is obtained from A by normalizing the rows corresponding to the tangible markings. The entire analysis can be carried on using Π and e alone; only at the end the nature of the markings is taken into account. After solving for γ^* in the equation $\gamma^* = \gamma^*\Pi$ subject to $\sum_{\mu \in \mathcal{R}} \gamma_\mu^* = 1$, we can obtain π^* using the following relation:

$$\forall \mu \in \mathcal{T}, \pi_\mu^* = \gamma_\mu^* h_\mu \cdot \left(\sum_{k \in \mathcal{T}} \gamma_k^* h_k\right)^{-1},$$

where h_k is the expected sojourn time for each visit to tangible marking k, that is, $h_k = -Q_{k,k}^{-1}$. For the visits to the transient states, it is more natural to compute n^* first, by solving $n^*(I - \overline{\Pi}) = \overline{e}$, where $\overline{\Pi}$ is the restriction of π to the transient states. Then:

$$\forall \mu \in \mathcal{T}, \sigma_\mu^* = n_\mu^* h_\mu.$$

Preservation is easier to describe and has useful properties (e.g., the maximum number of nonzero entries in any row of Π is bounded by $|T|$; this is not the case for Q), but the additional memory to store the vanishing markings often makes it an inferior choice. Also, we do not know of a way to perform transient analysis using Π instead of Q.

Ciardo et al. compared these approaches and discussed how to perform sensitivity analysis, that is, the computation of the derivatives of the measures of interest with respect to a real parameter affecting the rates or weights of any number of transitions[20]. The entire solution process has also been analyzed in detail[16]. For numerical solution methods of the underlying CTMC's, see Stewart's excellent book[52].

9.4. DISCRETE-TIME MARKOV SPN'S

The discrete analogous of the exponential distribution is the geometric distribution, which also enjoys the memoryless property, at integral instants of time. The initial definition of discrete-time SPN's was then

limited to geometrically distributed transitions firing times[42]. These
have been extended[18] to include immediate transitions, geometric firing
distributions with steps multiple of the basic unit step, and integral
constants (geometric distributions with success parameter equal one).

An even more general definition is possible[15], where firing time distributions are of the DDP type (defective discrete phase), corresponding to the time to reach a trap state 0 in an arbitrary finite DTMC
(the restriction to finite DTMC's is only a practical concern). These
are the distributions obtained from Const(0), Const(1), and Const(∞)
through the operations of finite convolution, finite probabilistic mixture, geometric sum, and finite order statistics.

The resulting DTMC-SPN's have, in principle, only two limitations.
The first one is the need to assume that all activities in the system to
be modeled have DDP distributions with a common unit step. Since
it is well known that any distribution can be approximated arbitrarily well by a (discrete or continuous) phase-type distribution, this only
means that we might be forced to use a fine step in practice. The
second limitation is the size of the state space, which grows both with
the structural complexity of the untimed Petri net (number of reachable markings) and with the number of phases used to describe the
distributions (a finer step implies a larger state space).

This is because the state of a DTMC-SPN is described by $s = (\mu, \psi)$,
where μ is the current marking and ψ is the current phase of each
transition's firing process. From now on, we use the symbol $\mathcal{R} \subseteq \mathbb{N}^{|P|}$
to indicate the set of reachable markings, and $\mathcal{S} \subseteq \mathbb{N}^{|P|} \times \mathbb{N}^{|T|}$ to
indicate the set of reachable states. Formally, a DTMC-SPN is a tuple
$\left(P, T, D^-, D^+, D^\circ, \succ, g, \mu^{[0]}, G, F, \psi^{[0]}, w \right)$ where:

- $\left(P, T, D^-, D^+, D^\circ, \succ, g, \mu^{[0]} \right)$ define an extended Petri net.

- $\forall \mu \in \mathbb{N}^{|P|}, \forall t \in \mathcal{E}(\mu), \forall i, j \in \mathbb{N}, G_t(\mu, i, j)$ is the probability that
  the phase of t changes from i to j at the end of one step, when t
  is enabled in marking μ.

- $\forall \mu \in \mathbb{N}^{|P|}, \forall u \in \mathcal{E}(\mu), \forall t \in T, \forall i, j \in \mathbb{N}, F_{u,t}(\mu, i, j)$ is the prob-
  ability that the phase of t changes from i to j when u fires in
  marking μ.

- $\forall t \in T, \psi_t^{[0]} \in \mathbb{N}$ is the phase of t at time 0.

- $\forall \mu \in \mathbb{N}^{|P|}, \forall S \subseteq \mathcal{E}(\mu), \forall t \in S, w_{t|S}(\mu) \geq 0$ is the firing weight for
  t when S is the set of *candidates* (to fire) in marking μ.

A transition $t \in T$ is said to be a candidate in state $s = (\mu, \psi)$ iff t is enabled in μ and $\psi_t = 0$. Let $C(s)$ be the set of candidates in state s. The probability that $t \in C(s)$ fires immediately next is

$$\hat{w}_{t|C(s)}(\mu) = w_{t|C(s)}(\mu) \cdot \left(\sum_{u \in C(s)} w_{u|C(s)}(\mu)\right)^{-1},$$

where we assume that the denominator is positive, that is, at least one candidate has nonzero weight whenever $|C(s)| > 0$. We call these states vanishing, while states with no candidates are tangible. This implies a "race policy[2]": only the enabled transitions whose firing time has elapsed can fire.

$G_t(\mu, \cdot, \cdot)$ is the one-step transition probability matrix for the DTMC $\{\psi_t^{[k]} : k \in \mathbb{N}\}$ describing the firing time distribution of t in marking μ in isolation. If another transition u fires before t, leading to marking μ', the phase ψ_t of t will change according to the distribution $F_{u,t}(\mu, \psi_t, \cdot)$ and, after that, it will evolve according to $G_t(\mu', \cdot, \cdot)$, which might differ from $G_t(\mu, \cdot, \cdot)$.

$G_t(\mu, \cdot, \cdot)$ allows us to model complex distributions. If there is no need for marking dependency in its specification, we can simply define it independent of μ. The reasons for defining $F_{u,t}(\mu, \psi_t, \cdot)$ are more complex. In the exponential case of GSPN's, we do not have to worry about what happens to the elapsed firing time of t when u fires: the memoryless property ensures that restarting or continuing it are probabilistically equivalent. However, that is not the case for DTMC-SPN's, where we have the following choices:

- If t is disabled by the firing of u, and we leave the phase ψ_t of t unchanged so that we continue its firing time when it becomes enabled again, we model "age memory[2]".

- If we resample ψ_t when t becomes enabled, we model "enabling memory[2]".

- We can model specialized a behavior where, for example, the phase of t is moved "closer" or "farther" to 0, depending on the marking μ, the current phase ψ_t, and u.

9.4.1. Solution algorithms

The DTMC underlying a DTMC-SPN is described by a transition probability matrix Π. First, we observe that a sequence of events can lead from tangible state $s = (\mu, \psi)$ to tangible state $s' = (\mu', \psi')$ in one timestep. This takes into account the evolution of the firing times in isolation for one timestep, from $s = (\mu, \psi)$ to $s^{(0)} = (\mu^{(0)}, \psi^{(0)})$, according to the G matrices, followed by what can happen at the end of this step in zero time. Once in state $s^{(i)} = (\mu^{(i)}, \psi^{(i)})$, if $C(s^{(i)})$ is not empty, one of them, $t^{(i)}$, will be chosen to fire, according to the weights. Its firing will cause a phase change for all transitions, according to the F matrices, and a marking change, resulting in the new state $s^{(i+1)} = (\mu^{(i+1)}, \psi^{(i+1)})$, and so on, until a tangible state $s^{(n)} = s'$ is reached. The contribution to $\Pi_{s,s'}$ for this sequence is

$$\left(\prod_{t \in \mathcal{E}(\mu)} G_t\left(\mu, \psi_t, \psi_t^{(0)}\right) \right) \prod_{i=0}^{n-1} \left(\hat{w}_{t^{(i)}|C(s^{(i)})}\left(\mu^{(i)}\right) \prod_{t \in T} F_{t^{(i)},t}\left(\mu^{(i)}, \psi_t^{(i)}, \psi_t^{(i+1)}\right) \right).$$

In a practical implementation, Π is computed one row at a time. The complexity of computing row s of Π can be substantial, depending on the length and number of sequences to be considered.

Just as in the CTMC-SPN's, we are interested in the accumulated or instantaneous reward processes $\{Y(\theta), \theta \geq 0\}$ and $\{y(\theta), \theta \geq 0\}$. If the DTMC has a finite state space, its analysis can be carried on in a similar way to that of CTMC-SPN's. Let $\pi^{[n]} = \left[\pi_s^{[n]}\right] = \left[\text{Prob}\{s^{[n]} = s\}\right]$ be the state probability vector at step n. Given the one-step transition probability matrix Π and the initial state probability vector $\pi^{[0]}$, the state probability vector at step n can be obtained using the *power method*, with the iteration: $\pi^{[n]} = \pi^{[n-1]}\Pi$. Since the DTMC can change state only at integer times, $\pi(\theta) = \pi^{[n]}$ for $\theta \in [n, n+1)$.

Assuming an ergodic DTMC, the steady-state probability vector π^* can be obtained as the solution of $\pi^* = \pi^*\Pi$, which can be rewritten as the homogeneous linear system $\pi^*(\Pi - I) = 0$, subject to $\sum_{s \in S} \pi_s^* = 1$. This is exactly the same type of equation as in Section 9.3.1, so the same solution methods can be employed.

We now consider the approximate computation of the expected value of $y(n_F)$ and $Y(n_F)$ for finite values of n_F. Let $S^{[n]}$ be the set of states having nonzero probability at step n. We can define $S^{[n]}$ inductively: $S^{[0]} = \{s \in S : \pi_s^{[0]} > 0\}$, and $S^{[n]} = \{s' \in S : \exists s \in S^{[n-1]}, \Pi_{s,s'} > 0\}$. We can then compute and store only the nonzero entries of $\pi^{[n]}$. These correspond to states in $S^{[n]}$, which is finite, regardless of whether S is finite or not, if:

- $n_F < \infty$, transient analysis,

- $|\{s : \pi_s^{[0]} > 0\}| < \infty$, a finite set of states has nonzero initial probability, and

- $\forall s \in \mathcal{R}, |\{s' : \Pi_{s,s'} > 0\}| < \infty$, each state can only reach a finite number of states in one step.

In practice, the third requirement might be hard to verify a priori. More importantly, though, $S^{[n]}$ could still be too large. We can then use a *truncation* approach, resulting in bounds on the measures of interest. This is applicable even when the above conditions are not satisfied, that is, even when some or all of the sets $S^{[n]}$ are infinite. The idea is based on deleting states with low probability, while keeping track of the probability mass unaccounted for. At step n, only the states in $\hat{S}^{[n]} \subseteq S^{[n]}$ are kept. For each state $s \in \hat{S}^{[n]}$, its computed probability $\hat{\pi}_s^n$ is an approximation of the exact probability $\pi_s^{[n]}$ at step n.

Step 1. If only a finite number of states have nonzero probability initially, set

$$\hat{\pi}^{[0]} \leftarrow \pi^{[0]} \quad \text{and} \quad \hat{S}^{[0]} \leftarrow \{s : \hat{\pi}_s^{[0]} > 0\}.$$

Otherwise, store in $\hat{S}^{[0]}$ only a finite subset of them, ideally chosen from $S^{[0]}$ in order of decreasing probability. The "total known probability mass" and the "total known sojourn time" at the beginning are

$$\kappa^{[0]} \leftarrow ||\hat{\pi}^{[0]}||_1 = \sum_{s \in \hat{S}^{[0]}} \hat{\pi}_s^{[0]} \leq 1 \quad \text{and} \quad K^{[0]} \leftarrow 0.$$

Step 2. Iterate using the power-method:

for each $s \in \hat{S}^{[n-1]}$ and $\Pi_{s,s'} > 0$ do $\quad \hat{\pi}_{s'}^{[n]} \leftarrow \hat{\pi}_{s'}^{[n]} + \hat{\pi}_s^{[n-1]} \Pi_{s,s'}$,

storing only the nonzero elements of the vector $\hat{\pi}^{[n]}$, which is initially empty, hence implicitly initialized to 0. If the number of states with nonzero probability becomes too large, truncate the states with probability below a threshold c, removing them from $\hat{\pi}^{[n]}$, de facto setting their probability to zero:

for each $s \in \hat{S}^{[n]}$ do $\quad$ if $\hat{\pi}_s^{[n]} < c$ then $\quad \hat{\pi}_s^{[n]} \leftarrow 0$.

Regardless of whether truncation is performed, the set of kept states is

$$\hat{S}^{[n]} \leftarrow \{s : \hat{\pi}_s^{[n]} > 0\},$$

the total known probability mass and sojourn time up to step n are

$$\kappa^{[n]} \leftarrow ||\hat{\pi}^{[n]}||_1 = \sum_{s \in \hat{S}^{[n]}} \hat{\pi}_s^{[n]} \leq 1 \quad \text{and} \quad K^{[n]} \leftarrow K^{[n-1]} + \kappa^{[n]} \leq n,$$

and the approximate value of the accumulated reward up to time n is

$$E[\hat{Y}(n)] \leftarrow E[\hat{Y}(n)] + \sum_{s \in \hat{S}^{[n-1]}} \hat{\pi}_s^{[n-1]} \rho(s).$$

Again, if an infinite number of states can be reached from a state s in a single timestep, generate only a finite number of them, ideally the ones corresponding to the largest entries in $\Pi_{s,\bullet}$.

Step 3. Upon reaching step n_F, the probability of being in state $s \in \hat{S}^{[n_F]}$ at step n is at least $\hat{\pi}_s^{[n_F]}$, but there is no way of knowing how the unaccounted probability mass $\kappa^{[0]} - \kappa^{[n_F]}$ should be redistributed (it should be redistributed over the states in $S^{[n_F]}$, hence some of it could be over states in $\hat{S}^{[n_F]} \subseteq S^{[n_F]}$). An analogous interpretation holds for $K^{[n_F]}$. Hence, assuming that the reward rates associated to the states have an upper and lower bound ρ_L and ρ_U, respectively, $E[Y(n_F)]$ and $E[y(n_F)]$ can be bounded as well. If $E[\hat{Y}(n_F)]$ and $E[\hat{y}(n_F)] = \sum_{s \in \hat{S}^{[n_F]}} \hat{\pi}_s^{[n_F]} \rho(s)$ are the approximations obtained using our truncation approach,

$$E[\hat{y}(n_F)] + \rho_L(1 - \kappa^{[n_F]}) \leq E[y(n_F)] \leq E[\hat{y}(n_F)] + \rho_U(1 - \kappa^{[n_F]}),$$

$$E[\hat{Y}(n_F)] + \rho_L(n_F - K^{[n_F]}) \leq E[Y(n_F)] \leq E[\hat{Y}(n_F)] + \rho_U(n_F - K^{[n_F]}).$$

Truncation can be performed as many times as needed, but every application reduces the quality of the bounds. Highly-reliable systems are particularly good candidates for this state-space truncation, since most of their states have an extremely low probability.

We conclude with a few observations. First, DDP distributions do not have a canonical representation and, unfortunately, the size of the DTMC underlying a DTMC-SPN may be affected by the choice of the representation for the DDP distributions involved[15]. Second, the "dynamic" state space exploration presented here has been suggested also in the CTMC case[26,30], although $S^{[n]}$ is a non-decreasing sequence of sets, in this case. Also, the uniformization approach for the solution of CTMC's follows a similar idea. However, for DTMC's, no truncation on the number of steps is required. The number of steps to be considered is determined by the time n_F at which the results are sought.

Returning to our example of Fig. 9.2, the use of an exponential distribution for transition *tout* is hardly defensible: a constant timing wold be much more appropriate. Hence, we could use the firing time distributions shown in Table 9.4 (DTMC-SPN case). We choose a geometric distribution (with unit step) as the default for a very practical

Table 9.4: Firing time distributions for the transmission protocol SPN.

SPN type	gen	$xmit_M$	rcv_M	$xmit_A$	rcv_A	tout
CTMC-SPN	Expo	Expo	Expo	Expo	Expo	Expo
DTMC-SPN	Geom	Geom	Geom	Geom	Geom	Const(n)
SMP-SPN	any	any	any	any	any	Expo
MRGP-SPN	any	any	Expo	Expo	Expo	Const(c)

reason: it corresponds to the smallest increase in the size of the state space while still being able to match the expected duration of the modeled activity, as long as it is at least one unit step: if $E[X] = x \geq 1$, we can approximate X with the distribution Geom(x^{-1}) to match its mean value. We reserve more complex distributions to activities we know have a highly non-memoryless behavior such as *tout*. The size of the tangible state-space S is then affected by the timeout duration, assumed to be $n \in \mathbb{N}$ unit steps. The underlying DTMC will have:

- One state for each of the markings 0 and 1 of Table 9.1.

- n, $n-1$, and $n-2$ states corresponding to the markings 2, 3, and 4, respectively, of Table 9.1. This is because, in marking 2, the phase of *tout* can be n, $n-1$, ...1. In marking 3, ψ_{tout} cannot be n, since at least one unit step is required for the token to go from $transit_M$ to *remote*, and so on.

Hence, the total number of reachable states is $2 + n + (n-1) + (n-2) = 3n - 1$, assuming $n \geq 2$. In practice, we would like to set the "unit step" so that n is a small integer, hence S is is not too large, but this is not possible in our case, since the timeout interval should be substantially larger than the expected amount of time required for a successful communication to complete, and this is the sum of the three expected times, for rcv_M, $xmit_A$, and rcv_A. Given the limitation that the parameter of the geometric distribution be no greater than one (and substantially less than one if we seek a "smooth" distribution), it is likely that n would have values in the range 10 — 100. This combinatorial increase in the state-space shows a limitation of the phase-type approach. Had we wanted to use a CTMC-SPN, we would have used an Erlang distribution for *tout* and an exponential distribution for the other transitions. Also in this case, though, the size of the state space would increase with the number of phases.

9.5. SEMI-MARKOV, REGENERATIVE SPN'S

If the firing distributions are allowed to be general, the state s of a SPN can still be described by two components. The first one is the marking μ, but the second one is now, in general, a continuous vector describing the *remaining firing times* (RFT's) τ of the transitions. Hence the state is $s = (\mu, \tau) \in \mathbb{N}^{|P|} \times (\mathbb{R}^0)^{|T|}$, where τ_t is the time that must elapse while transition t is enabled, before it can fire. This extension of the concept of phase from the previous section, however, makes the numerical solution intractable in general.

Under certain conditions, however, the underlying stochastic process is still solvable numerically. The easiest case occurs when the SPN defines a semi-Markov process (SMP). This happens if any number of transitions with arbitrary distributions can be enabled in a marking, but the firing of any of them resets the firing time of each nonexponentially distributed transition. This ensures that the RFT's in the new marking depend only on the new marking, not on the value of the the RFT's in the previous marking[18].

In our simple model of the transmission protocol, the underlying process is a SMP if the transitions have the firing time distributions shown in Table 9.4 (SMP-SPN case). Given the initial marking of Fig. 9.2, *gen* and $xmit_M$ are always enabled in isolation, hence their firing time can never be interrupted. Transitions rcv_M, $xmit_A$, and rcv_A, instead, are always enabled concurrently to *tout*, but they become disabled if *tout* fires, so no RFT needs to be retained for them when the marking changes. The SMP is shown in Fig. 9.5 (SMP-SPN case), where the transition names implicitly identify the corresponding distribution. The reason for requiring an exponential distribution for transition *tout* is apparent: it becomes enabled in marking 2, but it remains enabled as the marking changes to 3 or 4. Due to the memoryless property, however, the distribution of τ_{tout} after rcv_M or $xmit_A$ fire is the same as the distribution of the entire firing time for *tout*; we might treat it as if it were reset.

Unfortunately, few practical SPN's satisfy the restrictive conditions ensuring an underlying SMP. For example, in our model, transition *tout* is the least likely to have an exponential distribution, since resetting the timer after rcv_M or $xmit_A$ fires is exactly the opposite behavior one would expect. Indeed, it would be preferable to use deterministic timing for *tout* and random timing for the other transitions. Unless the phase-type approach of the previous section is used, the resulting process is not a SMP. However, it still enjoys a very important property:

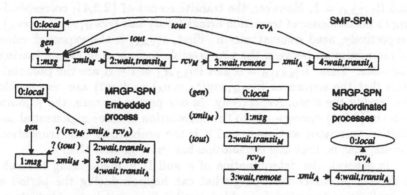

Figure 9.5: The underlying process for the transmission protocol.

its evolution contains a sequence of *regeneration points*. Every time transitions *gen*, $xmit_M$, rcv_A, or *tout* fire, the future evolution of the SPN is completely determined without making use of the past history. This is the Markov-regenerative property. The SPN is such that every firing leading to marking 0, 1, or 2 is a regeneration point, regardless of the firing time distributions.

Initial work on this class of SPN's assumed that all enabled transitions have exponential distributions except at most one enabled transition with deterministic[6,11,10] or general[12,18] distribution. We then illustrate the general idea first, assuming the distributions shown in Table 9.4 (MRGP-SPN case). Consider first the *embedded process* shown in Fig. 9.5 (MRGP-SPN case). This is a SMP with state space $\mathcal{E}$. It is well known that the steady-state probability of its states can be computed from the expected sojourn times for each visit to state $i \in \mathcal{E}$, $[\sigma_i]$, and from the transition probability matrix, $[\Pi_{i,j}]$, describing the probability that the next state visited is j, given that the state just entered is i (we assume this describes an ergodic DTMC).

First, we compute the steady-state solution γ^* of the embedded DTMC, solving $\gamma^* = \Pi \gamma^*$ subject to $\sum_{j \in \mathcal{E}} \gamma_j^* = 1$, then, we compute the steady-state solution of the SMP using the sojourn times as weights:

$$\forall i \in \mathcal{E}, \ \pi_i^* = \gamma_i^* \sigma_i \cdot \left(\sum_{j \in \mathcal{E}} \gamma_j^* \sigma_j \right)^{-1}.$$

The transitions from 0 to 1 and from 1 to the *macro state* $\{2,3,4\}$ are governed by the distributions of *gen* and $xmit_M$, respectively. Hence, σ_0 and σ_1 are simply the expected values of these distributions, $\Pi_{0,1} = 1$, and $\Pi_{1,\{2,3,4\}} = 1$. However, the transitions out of $\{2,3,4\}$, corresponding to the sequence of transition firings $(tout)$ and $(rcv_M, xmit_A, rcv_A)$, respectively, need some attention. First, $\sigma_{\{2,3,4\}}$ is the expected value of the minimum between the two possible sequences of events leaving the state. Then, $\Pi_{\{2,3,4\},0} = \alpha$ and $\Pi_{\{2,3,4\},1} = 1 - \alpha$ are the probabilities that the sequences $(rcv_M, xmit_A, rcv_A)$ or $(tout)$ are responsible for leaving the state, respectively. In our particular case, the sojourn time in $\{2,3,4\}$ corresponds to the convolution of three exponential activities, truncated at time c, and α is the probability that truncation occurs, that is, that this convolution has value greater than c.

In general, the determination of σ and Π is made using the *subordinated processes* describing what can happen during the period a *subordinating transition* is enabled, until it fires or it is disabled, starting from a particular marking in which it becomes enabled. Fig. 9.5 shows three subordinated processes, corresponding to the enabling of the subordinating transitions gen, $xmit_M$, and $tout$, respectively. The firing of the subordinating transition is not explicitly represented, it can occur in each state drawn with a rectangular shape. The first two processes are trivial: no other transition is enabled, so the subordinating transitions eventually fire, according to their firing time in isolation. The third process, however, considers the evolution of the SPN from the instant $tout$ becomes enabled, in marking 2. Marking 0 is drawn with an oval shape because the subordinating transition $tout$ is disabled upon entering it.

A subordinated process is used to compute:

- For each state s, the probability π'_s of being in it at the time the subordinating transition would fire in isolation. These probabilities are then used to compute Π. In the case of the third subordinated process, for example, we obtain that $\Pi_{\{2,3,4\},0} = \pi'_0$ and $\Pi_{\{2,3,4\},1} = \pi'_2 + \pi'_3 + \pi'_4$.

- For each state s enabling the subordinating transition, the cumulative amount of time σ'_s spent in it up to the time the subordinating transition would fire in isolation. These are used to compute σ. For example, $\sigma_{\{2,3,4\}} = \sigma'_2 + \sigma'_3 + \sigma'_4$.

Once the embedded process is solved, the probability of being in a macro state is split over its component states proportionally to σ':

$$\pi_2^* = \pi_{\{2,3,4\}}^* \frac{\sigma'_2}{\sigma_{\{2,3,4\}}} \qquad \pi_3^* = \pi_{\{2,3,4\}}^* \frac{\sigma'_3}{\sigma_{\{2,3,4\}}} \qquad \pi_4^* = \pi_{\{2,3,4\}}^* \frac{\sigma'_4}{\sigma_{\{2,3,4\}}}.$$

If all the transitions in the subordinated process are exponentially distributed and if the subordinating transition has a constant firing time equal c, the solution of the subordinated process simply requires the transient analysis of a CTMC at time c. This is the case of the "deterministic and stochastic Petri nets" (DSPN's)[6,11,19]. If the same subordinating transition can become enabled in different markings, multiple subordinated processes must be considered, and the same state may belong to multiple macro states.

It also possible to compute "up-to-absorption" type of measures, if the embedded process contains transient states[19]. A simple scaling can be used to model marking-dependent deterministic firing times[28,40].

If the distribution of *tout* were not constant, the time at which we need to study the subordinated process is a random variable. When the firing time of *tout* has an expolynomial distribution (a quite general class), the approach used for DSPN's is still applicable with small modifications, and it retains the same complexity[12,18].

9.6. FURTHER ISSUES

We now mention important additional directions of SPN research. The interested reader can use the references listed for further information.

9.6.1. Well-defined SPN's

Contemporary firings are a frequent source of errors when specifying a SPN, since the order in which they are sequentialized might affect the behavior (*stochastic confusion*). This happens regularly in the DTMC-SPN's, but the problem already arises in the GSPN's.

It is then essential to ensure that the SPN is "well defined", that is, its stochastic behavior is completely determined. This can be achieved in two ways: either at the net level, restricting the type of SPN's that can be defined[9], or at the reachability graph or stochastic process level, ensuring that, as the analysis proceeds, the parameters of the probabilistic evolution of the SPN can be determined[23,48]. These approaches are normally described in relation to an analytical solution, but they can (and should) be applied to simulation as well.

9.6.2. Stochastic well-formed colored nets

Many extensions using *colored tokens* have been proposed[36,38], where a transition is enabled only for given combinations of token colors in its input places. This is often essential to describe complex behaviors in a compact way.

Considering our protocol SPN, we might want to allow up to $K \geq 1$ outstanding (not yet acknowledged) messages. This can be modeled with K tokens initially in place *local*. However, consider now the marking where there are k_1, k_2, and k_3 tokens in places $transit_M$, *remote*, and $transit_A$, respectively (hence $k = k_1 + k_2 + k_3 \leq K$ tokens in place *wait*). If transition *tout* fires, it means that one of the k concurrent timers has elapsed, and we should remove the corresponding token from one of $transit_M$, *remote*, or $transit_A$. The cardinalities of the input arcs from these three places should be still defined so that one of them evaluates to one and the others to zero, but the model does not contain enough information to choose the correct one.

We then have two options. We can modify the SPN appropriately, or we can use tokens with K different colors, so that the color of the token removed from *wait* when *tout* fires matches the color of the token removed from $transit_M$, *remote*, or $transit_A$.

Initially, the use of colors was considered a mere convenience, but the solution methods were based either on simulation, or on a *decolorization* of the SPN. In the latter case, the state space increases dramatically. It is with this problem in mind that the stochastic well-formed colored nets (SWN's) were defined[10]. By limiting the rules for the use of colors, the SWN formalism allows to build a *symbolic reachability graph* without decolorizing the net. For example, in our (now colored) protocol SPN, we defined K colors, but any marking with k_1, k_2, and k_3 tokens in $transit_M$, *remote*, and $transit_A$ can be considered equivalent, regardless of the token colors (this is of course true only if the distributions are exponential and the timing behavior of the tokens is independent of the color). With SWN's, it is possible to recognize this situation and exploit the symmetries automatically.

9.6.3. Product-form SPN's

Given the combinatorial growth of the state space, it would be desirable to derive analytical algorithms to compute measures defined on the SPN without having to generate the state space. Researchers have been looking for a SPN counterpart to the product-form queuing networks, which made a profound practical impact in the system performance evaluation world.

In the last few years, substantial advances have been made. First, a condition for a product-form solution was defined[33]. Then, several algorithms for the solution were given[24,51]. Unfortunately few SPN's exhibit (exact) product form characteristics, but approximate algorithms have been recently proposed[50].

9.6.4. Approaches based on Kronecker algebra

An alternative class of approaches to achieve a more efficient SPN solution has been based, loosely speaking, on the idea of near-independence: different parts of a system usually evolve fairly independently of each other, with only occasional interactions.

Building on Plateau's work on stochastic automata[47], Donatelli[27], Buchholz[7,8], and Kemper[39] have defined increasingly complex classes of GSPN's which can be described as the composition of smaller models. The key advantage of this approach is that the overall infinitesimal generator matrix Q can be defined in terms of much smaller matrices $Q_1, Q_2, \ldots Q_k$ and some corrective matrices, combined using Kronecker sums or products. Thus, Q itself, the largest data structure needed for the solution, is not stored explicitly. A recent development is the application of these ideas to SWN's[32].

In the approximate area, related approaches exploit the idea of fixed-point iteration, often using heuristics, to decompose a SPN into submodels. Their advantage is the large reduction in the solution time, but no estimation of the error is available, as is often the case for these iterative approaches [17,22]. Existence[41] and, even more, uniqueness of a fixed point guaranteeing convergence to a single solution independent of the initial guess are also hard to prove.

9.7. CONCLUSION

We have attempted to present an overview of the theory behind various classes of SPN's and their applicability to modeling practical problems. Special emphasis has been placed on the numerical methods used when the underlying process described by the SPN is Markov, semi-Markov, or Markov-regenerative.

For up-to-date information on this topic, the interested reader should consult, in addition to the standard computer science journals, the proceedings of the Petri Net and Performance Models (PNPM) Workshop, published by IEEE Computer Society Press, and the International Conference on Application and Theory of Petri Nets, published by Springer-Verlag in the Lecture Notes in Computer Science series. Extensive Petri net information is also maintained on the World Wide

Web site http://www.daimi.aau.dk/~petrinet/, including a list of software tools for analyzing untimed, timed, and stochastic Petri nets.

REFERENCES

[1] AJMONE MARSAN, M., BALBO, G., BOBBIO, A., CHIOLA, G., CONTE, G., AND CUMANI, A. On Petri nets with stochastic timing. In *Proc. Int. Workshop on Timed Petri Nets* (Torino, Italy, July 1985), pp. 80–87.

[2] AJMONE MARSAN, M., BALBO, G., BOBBIO, A., CHIOLA, G., CONTE, G., AND CUMANI, A. The effect of execution policies on the semantics and analyis of Stochastic Petri Nets. *IEEE Trans. Softw. Eng. 15*, 7 (July 1989), 832–846.

[3] AJMONE MARSAN, M., BALBO, G., CHIOLA, G., AND CONTE, G. Generalized Stochastic Petri Nets revisited: random switches and priorities. In *Proc. 2nd Int. Workshop on Petri Nets and Performance Models (PNPM'87)* (Madison, Wisconsin, Aug. 1987), IEEE Comp. Soc. Press, pp. 44–53.

[4] AJMONE MARSAN, M., BALBO, G., AND CONTE, G. A class of Generalized Stochastic Petri Nets for the performance evaluation of multiprocessor systems. *ACM Trans. Comp. Syst. 2*, 2 (May 1984), 93–122.

[5] AJMONE MARSAN, M., BALBO, G., CONTE, G., DONATELLI, S., AND FRANCESCHINIS, G. *Modelling with generalized stochastic Petri nets.* John Wiley & Sons, 1995.

[6] AJMONE MARSAN, M., AND CHIOLA, G. On Petri nets with deterministic and exponentially distributed firing times. In *Adv. in Petri Nets 1987, Lecture Notes in Computer Science 266*, G. Rozenberg, Ed. Springer-Verlag, 1987, pp. 132–145.

[7] BUCHHOLZ, P. Numerical solution methods based on structured descriptions of Markovian models. In *Computer performance evaluation* (1991), G. Balbo and G. Serazzi, Eds., Elsevier Science Publishers B.V. (North-Holland), pp. 251–267.

[8] BUCHHOLZ, P., AND KEMPER, P. Numerical analysis of stochastic marked graphs. In *Proc. Int. Workshop on Petri Nets and Performance Models (PNPM'95)* (Durham, NC, Oct. 1995), IEEE Comp. Soc. Press, pp. 32–41.

[9] CHIOLA, G., AJMONE MARSAN, M., BALBO, G., AND CONTE, G. Generalized Stochastic Petri Nets: a definition at the net level and its implications. *IEEE Trans. Softw. Eng. 19*, 2 (Feb. 1993), 89–107.

[10] CHIOLA, G., DUTHEILLET, C., FRANCESCHINIS, G., AND HADDAD, S. Stochastic well-formed colored nets and symmetric modeling applications. *IEEE Trans. Comp. 42*, 11 (Nov. 1993), 1343–1360.

[11] CHOI, H., KULKARNI, V. G., AND TRIVEDI, K. S. Transient analysis of deterministic and stochastic Petri nets. In *Application and Theory of Petri Nets 1993, Lecture Notes in Computer Science*, M. Ajmone Marsan, Ed., vol. 691. Springer-Verlag, 1993, pp. 166–185.

[12] CHOI, H., KULKARNI, V. G., AND TRIVEDI, K. S. Markov regenerative stochastic Petri nets. *Perf. Eval. 20*, 1-3 (1994), 337–357.

[13] CIARDO, G. *Analysis of large stochastic Petri net models*. PhD thesis, Duke University, Durham, NC, 1989.

[14] CIARDO, G. Petri nets with marking-dependent arc multiplicity: properties and analysis. In *Application and Theory of Petri Nets 1994, Lecture Notes in Computer Science 815 (Proc. 15th Int. Conf. on Applications and Theory of Petri Nets, Zaragoza, Spain)* (June 1994), R. Valette, Ed., Springer-Verlag, pp. 179–198.

[15] CIARDO, G. Discrete-time Markovian stochastic Petri nets. In *Numerical Solution of Markov Chains '95* (Raleigh, NC, Jan. 1995), W. J. Stewart, Ed., pp. 339–358.

[16] CIARDO, G., BLAKEMORE, A., CHIMENTO, P. F. J., MUPPALA, J. K., AND TRIVEDI, K. S. Automated generation and analysis of Markov reward models using Stochastic Reward Nets. In *Linear Algebra, Markov Chains, and Queueing Models*, C. Meyer and R. J. Plemmons, Eds., vol. 48 of *IMA Volumes in Mathematics and its Applications*. Springer-Verlag, 1993, pp. 145–191.

[17] CIARDO, G., CHERKASOVA, L., KOTOV, V., AND ROKICKI, T. Modeling a scalable high-speed interconnect with stochastic Petri nets. In *Proc. Int. Workshop on Petri Nets and Performance Models (PNPM'95)* (Durham, NC, Oct. 1995), IEEE Comp. Soc. Press, pp. 83–92.

[18] CIARDO, G., GERMAN, R., AND LINDEMANN, C. A characterization of the stochastic process underlying a stochastic Petri net. *IEEE Trans. Softw. Eng. 20*, 7 (July 1994), 506–515.

[19] CIARDO, G., AND LINDEMANN, C. Analysis of deterministic and stochastic Petri nets. In *Proc. 5th Int. Workshop on Petri Nets and Performance Models (PNPM'93)* (Toulouse, France, Oct. 1993), IEEE Comp. Soc. Press, pp. 160–169.

[20] CIARDO, G., MUPPALA, J. K., AND TRIVEDI, K. S. On the solution of GSPN reward models. *Perf. Eval. 12*, 4 (1991), 237–253.

[21] CIARDO, G., MUPPALA, J. K., AND TRIVEDI, K. S. Analyzing concurrent and fault-tolerant software using stochastic Petri nets. *J. Par. and Distr. Comp. 15*, 3 (July 1992), 255–269.

[22] CIARDO, G., AND TRIVEDI, K. S. A decomposition approach for stochastic reward net models. *Perf. Eval. 18*, 1 (1993), 37–59.

[23] CIARDO, G., AND ZIJAL, R. Well-defined stochastic Petri nets. In *Proc. 4th Int. Workshop on Modeling, Analysis and Simulation of Computer and Telecommunication Systems (MASCOTS'96)* (San Jose, CA, USA, Feb. 1996), IEEE Comp. Soc. Press, pp. 278–284.

[24] COLEMAN, J. L. Algorithms for product-form stochastic Petri nets – A new approach. In *Proc. 5th Int. Workshop on Petri Nets and Performance Models (PNPM'93)* (Toulouse, France, Oct. 1993), IEEE Comp. Soc. Press, pp. 108–116.

[25] COLOM, J. M., AND SILVA, M. Convex geometry and semiflows in P/T nets. A comparative study of algorithms for the computation of minimal p-semiflows. In *10th Int. Conf. on Application and Theory of Petri Nets* (Bonn, Germany, 1989), pp. 74–95.

[26] DE SOUZA E SILVA, E., AND MEJÍA OCHOA, P. State space exploration in Markov models. In *Proc. 1992 ACM SIGMETRICS Conf. on Measurement and Modeling of Computer Systems* (Newport, RI, USA, June 1992), pp. 152–166.

[27] DONATELLI, S. Superposed generalized stochastic Petri nets: definition and efficient solution. In *Application and Theory of Petri Nets 1994, Lecture Notes in Computer Science 815 (Proc. 15th Int. Conf. on Applications and Theory of Petri Nets, Zaragoza, Spain)* (June 1994), R. Valette, Ed., Springer-Verlag, pp. 258–277.

[28] GERMAN, R. *Analysis of stochastic Petri nets with non-exponentially distributed firing times*. PhD thesis, Technical University of Berlin, Berlin, Germany, 1994.

[29] GRASSMANN, W. K. Means and variances of time averages in Markovian environments. *Eur. J. Oper. Res. 31*, 1 (1987), 132–139.

[30] GRASSMANN, W. K. Finding transient solutions in Markovian event systems through randomization. In *Numerical Solution of Markov Chains*, W. J. Stewart, Ed. Marcel Dekker, Inc., New York, NY, 1991, pp. 357–371.

[31] GRASSMANN, W. K., AND WANG, Y. Immediate events in Markov chains. In *Numerical Solution of Markov Chains '95* (Raleigh, NC, Jan. 1995), W. J. Stewart, Ed., pp. 163–176.

[32] HADDAD, S., AND MOREAUX, P. Evaluation of high-level Petri nets by means of aggregation and decomposition. In *Proc. Int. Workshop on Petri Nets and Performance Models (PNPM'95)* (Durham, NC, Oct. 1995), IEEE Comp. Soc. Press, pp. 11–20.

[33] HENDERSON, W., AND TAYLOR, P. G. Embedded processes in stochastic Petri nets. *IEEE Trans. Softw. Eng. 17*, 4 (Feb. 1991), 108–116.

[34] HORTON, G., AND LEUTENEGGER, S. T. A multi-level solution algorithm for steady state Markov chains. In *Proc. 1994 ACM SIG-METRICS Conf. on Measurement and Modeling of Computer Systems* (Nashville, TN, May 1994), pp. 191–200.

[35] HOWARD, R. A. *Dynamic Probabilistic Systems, Volume II: Semi-Markov and Decision Processes*. John Wiley and Sons, 1971.

[36] HUBER, P., JENSEN, K., AND SHAPIRO, R. M. Hierarchies in coloured Petri nets. In *Proc. 10th Int. Conf. on Application and Theory of Petri Nets* (Bonn, Germany, June 1989), pp. 192–209.

[37] JENSEN, K. Coloured Petri nets and the invariant method. *Theoretical Computer Science 14* (1981), 317–336.

[38] JENSEN, K. Coloured Petri nets. In *Petri Nets: Central Models and Their Properties, Lecture Notes in Computer Science 254* (1987), Springer-Verlag, pp. 248–299.

[39] KEMPER, P. Numerical analysis of superposed GSPNs. In *Proc. Int. Workshop on Petri Nets and Performance Models (PNPM'95)* (Durham, NC, Oct. 1995), IEEE Comp. Soc. Press, pp. 52–61.

[40] LINDEMANN, C., AND GERMAN, R. Modeling discrete event systems with state-dependent deterministic service times. *Discrete Event Dynamic Systems: Theory and Applications 3* (July 1993), 249–270.

[41] MAINKAR, V., AND TRIVEDI, K. S. Fixed point iteration using Stochastic Reward Nets. In *Proc. Int. Workshop on Petri Nets and Performance Models (PNPM'95)* (Durham, NC, Oct. 1995), IEEE Comp. Soc. Press, pp. 21–30.

[42] MOLLOY, M. K. *On the integration of delay and throughput measures in distributed processing models*. PhD thesis, UCLA, Los Angeles, CA, 1981.

234 G. CIARDO

[43] MURATA, T. Petri Nets: properties, analysis and applications. *Proc. of the IEEE 77*, 4 (Apr. 1989), 541–579.

[44] NATKIN, S. Reseaux de Petri stochastiques. These de docteur ingeneur, CNAM-Paris, Paris, France, June 1980.

[45] PETERSON, J. L. *Petri Net Theory and the Modeling of Systems.* Prentice-Hall, 1981.

[46] PETRI, C. *Kommunikation mit Automaten.* PhD thesis, University of Bonn, Bonn, West Germany, 1962.

[47] PLATEAU, B., AND ATIF, K. Stochastic Automata Network for modeling parallel systems. *IEEE Trans. Softw. Eng. 17*, 10 (Oct. 1991), 1093–1108.

[48] QURESHI, M. A., SANDERS, W. H., VAN MORSEL, A. P. A., AND GERMAN, R. Algorithms for the generation of state-level representations of stochastic activity networks with general reward structures. In *Proc. Int. Workshop on Petri Nets and Performance Models (PNPM'95)* (Durham, NC, Oct. 1995), IEEE Comp. Soc. Press, pp. 180–190.

[49] REISIG, W. *Petri Nets*, vol. 4 of *EATC Monographs on Theoretical Computer Science.* Springer-Verlag, 1985.

[50] SERENO, M. Approximate mean value analysis technique for non-product form solution stochastic Petri nets: an application to stochastic marked graphs. In *Proc. Int. Workshop on Petri Nets and Performance Models (PNPM'95)* (Durham, NC, Oct. 1995), IEEE Comp. Soc. Press, pp. 42–51.

[51] SERENO, M., AND BALBO, G. Computational algorithms for product form solution stochastic Petri nets. In *Proc. 5th Int. Workshop on Petri Nets and Performance Models (PNPM'93)* (Toulouse, France, Oct. 1993), IEEE Comp. Soc. Press, pp. 98–107.

[52] STEWART, W. J. *Introduction to the Numerical Solution of Markov Chains.* Princeton University Press, 1994.

[53] STOTTS, D. P., AND GODFREY, P. Place/transition nets with debit arcs. *Information Processing Letters 41* (Jan. 1992), 25–33.

[54] VALK, R. On the computational power of extended Petri nets. In *7th Symp. on Mathematical Foundations of Computer Science, Lecture Notes in Computer Science 64.* Springer-Verlag, 1978, pp. 527–535.

CHAPTER 10

STOCHASTIC PROCESS ALGEBRAS: A NEW APPROACH TO PERFORMANCE MODELING

Jane Hillston and Marina Ribaudo

10.1 INTRODUCTION

In this chapter we present an introduction to the novel approach to performance modeling provided by Stochastic Process Algebras (SPA). Like queueing networks and stochastic Petri nets, and their variants, these formal languages can be regarded as high-level model specification languages for low-level stochastic models. The advantages of SPAs are that they incorporate the attractive features of process algebras and thus bring to the area of performance modeling several attributes which are not offered by the existing formalisms. Perhaps the most important such feature is the compositionality which is inherent in the models and can be exploited during their analysis. Throughout this chapter we will present several simple examples which illustrate this and several other aspects of the SPA approach to performance modeling.

In recent years several SPA languages have been developed. Here we will focus upon Hillston's Performance Evaluation Process Algebra (PEPA). The other published languages are discussed in Section 10.6.

The rest of the chapter is structured as follows. In Section 10.2 we present a brief overview of process algebras and their recent timed and probabilistic extensions. PEPA is introduced in some detail in Section 10.3, and the associated approach to model construction is discussed.

235

The analysis of models and the extraction of performance measures are presented in Section 10.4. Facilities for manipulating models, model simplification and aggregation, are outlined in Section 10.5. In particular we discuss the advantages brought to such manipulation by the compositional structure and equivalence relations inherent in SPA. In Section 10.6 we give a summary of the different features of the other SPA languages and in Section 10.7 we outline recent research work which aims to exploit the compositional structure within SPA models to find efficient solution techniques. Finally we summarise the attractive features offered by this new approach.

10.2   CLASSICAL PROCESS ALGEBRA

Process algebras are abstract languages used for the specification and design of concurrent systems. The most widely used process algebras are Milner's Calculus of Communicating Systems (CCS)[20] and Hoare's Communicating Sequential Processes (CSP)[18] and the SPAs take inspiration from both these formalisms. Models in CCS and CSP have been used extensively to establish the correct behavior of complex systems by deriving *qualitative* properties such as *freedom from deadlock* or *livelock*.

In the process algebra approach systems are modeled as collections of entities, called *agents*, which execute atomic *actions*. These actions are the building blocks of the language and they are used to describe sequential behaviors which may run concurrently, and synchronisations or communications between them.

In CCS two agents communicate when one performs an action, a say, while the other performs the complementary action $\bar{a}$. The resulting communication action has the distinguished label τ, which represents an *internal* action that is invisible to the environment. Agents may proceed with their internal actions simultaneously but it is important to note that the semantics given to the language imposes an interleaving on such concurrent behavior. The grammar of the language makes it possible to construct an agent which has a designated first action (prefix); has a choice over alternatives (choice); or has concurrent possibilities (composition).

The communication mechanism is different in CSP as there is no notion of complementary actions: this is a major distinction between CCS and CSP. In CSP two agents communicate by simultaneously executing actions with the same label. Since during the communication the joint action remains visible to the environment, it can be reused by other concurrent processes so that more than two processes can be involved in the communication. This is the communication mechanism adopted in the SPA languages.

In CCS and CSP, since the objective is qualitative analysis rather than quantitative, time is abstracted away. In the last decade various suggestions for incorporating time into these formalisms have been investigated (for an overview see 21). However, most of these retain the assumption that actions are instantaneous and regard time progressing as orthogonal to the activity of the system. These assumptions make such models unsuitable for performance analysis. In contrast, SPAs associate a random variable, representing duration, with each action.

Similarly, process algebras are often used to model systems in which there is uncertainty about the behavior of a component, but this uncertainty is also abstracted away so that all choices become nondeterministic. Probabilistic extensions of process algebras, such as PCCS[19], allow this uncertainty to be quantified using a probabilistic choice combinator. In this case a probability is associated with each possible outcome of a choice. In SPA an alternative approach is taken—we assume that a *race condition* resolves choices when more than one action can occur.

10.3 PEPA

Process algebras offer several attractive features which are not necessarily available in existing performance modeling paradigms. The most important of these are *compositionality*, the ability to model a system as the interaction of its subsystems, *formality*, giving a precise meaning to all terms in the language, and *abstraction*, the ability to build up complex models from detailed components but disregarding internal behavior when it is appropriate to do so. Queueing networks offer compositionality but not formality; stochastic extensions of Petri nets offer formality but not compositionality; neither offer abstraction mechanisms.

Performance Evaluation Process Algebra[14] (PEPA) extends classical process algebra by associating a random variable, representing duration, with every action. These random variables are assumed to be exponentially distributed and this leads to a clear relationship between the process algebra model and a continuous time Markov chain (CTMC). Via this underlying CTMC performance measures can be extracted from the model.

PEPA models are described as interactions of *components*. Each component can perform a set of actions: an action $a \in \mathcal{A}ct$ is described as a pair (α, r), where $\alpha \in \mathcal{A}$ is the *type* of the action and $r \in \mathbb{R}^+$ is the parameter of the negative exponential distribution governing its duration. Whenever a process P can perform an action, an instance of a given probability distribution is sampled: the resulting number specifies how long it will take to *complete* the action. A small but powerful set of combinators is used to build up complex behavior from simpler behavior. The combinators are familiar: sequential composition, selection, synchronisation and encapsulation. We will explain each in its turn below. A formal operational semantics for PEPA is available in 14 and is not reproduced here.

Sequential composition: A component may have purely sequential behavior, repeatedly undertaking one activity after another and eventually returning to the beginning of its behavior. A simple example is a memory module in a multiprocessor system, which allows one data transfer at a time. Each processor requiring memory access will need to acquire the memory and only when the transfer is complete will the memory be released and available again for acquisition.

$$Mem \stackrel{\text{def}}{=} (get, \top).(use, \mu).(rel, \top).Mem$$

In some cases, as here, the rate of an action is outside the control of this component. Such actions are carried out jointly with another component, with this component playing a passive role. For example, the memory is passive with respect to the *get* action and this is recorded by the distinguished symbol, $\top$ (called "top").

Selection: A choice between two possible behaviors is represented as the sum of the possibilities. For example, if we consider a processor in the multiprocessor system, a computation may have two possible outcomes: access to local memory is necessary (with probability p_1) or access to a common memory module is necessary (with probability $p_2 = 1 - p_1$). These alternatives are represented as shown below:

$$\begin{aligned} Proc \stackrel{\text{def}}{=} \ &(think, p_1\lambda).(local, m).Proc \ + \\ &(think, p_2\lambda).(get, g).(use, \mu).(rel, r).Proc \end{aligned}$$

A race condition is assumed to govern the behavior of simultaneously enabled actions so the choice combinator represents pre-emptive selection with re-sampling. The continuous nature of the probability distributions ensures that the actions cannot occur simultaneously. Thus a sum will behave as either one summand or the other. When an action has more than one possible outcome, e.g. the *think* action in the processor, it is represented by a choice of separate actions, one for each possible outcome. The rates of these actions are chosen to reflect their relative probabilities.

Synchronisation: We have already anticipated that the processor and the memory in the example will be working together within the same system. This will require them to *cooperate* when the processor needs access to data which is not available locally. In contrast, the local activities of the processor can be carried out independently of the memory. Cooperation over given actions is reflected in the cooperation by the *cooperation set*, $L = \{get, use, rel\}$ in this case. Actions in this set require the simultaneous involvement of both components. The resulting action, a *shared* action, will have the same type as the two contributory actions and a rate reflecting the rate of the action in the slowest participating component. Note that this means that the rate of a passive action will become the rate of the action it cooperates with.

If, for simplicity, we assume that the multiprocessor consists of just two processors, the system is represented as the cooperation of the processors and the memory as follows:

$$Sys_1 \stackrel{\text{def}}{=} (Proc \parallel Proc) \bowtie_L Mem \qquad L = \{get, use, rel\}$$

The combinator $\parallel$ is a degenerate form of the cooperation combinator, formed when two components behave completely independently, without any cooperation between them, as in the case of the two processors. This pure parallel combinator can be thought of as cooperation over the empty set: $(Proc \bowtie_{\bullet} Proc)$.

Encapsulation: It is often convenient to hide some actions, making them private to the component or components involved. The duration of the actions is unaffected, but their type becomes hidden, appearing instead as the unknown type τ. Components cannot synchronise on τ. For example, as we further develop the model of the multiprocessor we may wish to hide the access of a processor to its local memory. This might lead to a new representation of the processor:

$$Proc' \stackrel{\text{def}}{=} Proc/\{local\}$$

and a corresponding new representation of the system:

$$Sys_2 \stackrel{\text{def}}{=} (Proc' \parallel Proc') \bowtie_L Mem \qquad L = \{get, use, rel\}$$

Use of the hiding combinator in this way has two implications. Firstly, it ensures that no components added to the model at a later stage can interact, or interfere, with this action of the processor. Secondly, as we will see in Section 10.4., private actions are deemed to have no contribution to the performance measures being calculated and this might subsequently suggest simplifications to the model.

Throughout the simple example above we have used constants such as Mem to associate names with behaviors. Using recursive definitions we have been able to describe components with infinite behaviors without the use of an explicit recursion operator.

Representing the components of the system as separate components means that we can easily extend our model. Now we may want to consider a system consisting of more than two processors which act independently of each other but compete for the use of common memory. This extension may be achieved compositionally by combining more instances of the $Proc$ component already described. For example, in the case of four processors we have:

$$Sys_3 \stackrel{\text{def}}{=} (Proc \parallel Proc \parallel Proc \parallel Proc) \bowtie_L Mem \qquad L = \{get, use, rel\}$$

10.4   MODEL ANALYSIS

The formality of the process algebra approach allows us to assign a precise meaning to every language expression. This implies that once we have a language description of a given system its behavior can be deduced automatically. The meaning, or semantics, of a PEPA expression is provided by the formal semantics, in the structured operational style, which associates a labeled multi-transition system with every expression in the language[14].

A labeled transition system $(S, T, \{\xrightarrow{t} \mid t \in T\})$ consists of a set of states S, a set of transition labels T and a transition relation

$\xrightarrow{t} \subseteq S \times S$. For PEPA the states are the syntactic terms in the language, the transition labels are the actions $((type, rate)$ pairs), and the transition relation is given by the semantic rules. A multi-transition relation is used because the number of instances of a transition (action) is significant since it can affect the timing behavior of a component.

Based on the transition relation, a transition diagram, called the *derivation graph* (DG), can be associated with each language expression. This graph describes all the possible evolutions of any component and provides a useful way to reason about the behavior of a model. An example derivation graph is shown in Figure 10.1 where the DG of the PEPA model Sys_4, consisting of a single processor accessing the global memory, is shown. For simplicity, in the upper part of the figure we have expanded the derivatives of the components $Proc$ and Mem.

$Proc \stackrel{\text{def}}{=} (think, p_1\lambda).Proc_1 + (think, p_2\lambda).Proc_2$

$Proc_1 \stackrel{\text{def}}{=} (local, m).Proc$

$Proc_2 \stackrel{\text{def}}{=} (get, g).Proc_3 \qquad Proc_3 \stackrel{\text{def}}{=} (use, \mu).Proc_4 \qquad Proc_4 \stackrel{\text{def}}{=} (rel, r).Proc$

$Mem \stackrel{\text{def}}{=} (get, \top).Mem_1 \qquad Mem_1 \stackrel{\text{def}}{=} (use, \mu).Mem_2 \qquad Mem_2 \stackrel{\text{def}}{=} (rel, \top).Mem$

$Sys_4 \stackrel{\text{def}}{=} Proc \underset{L}{\bowtie} Mem \qquad L = \{get, rel, use\}$

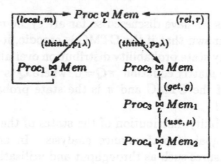

Figure 10.1: Derivation graph underlying Sys_4.

Inspection of the DG allows one to derive qualitative properties of the model. In this case, for instance, we can see that the PEPA model is free from deadlock and live. Moreover, the CTMC underlying any finite PEPA component can be obtained directly from the DG: a state of the Markov process is associated with each node of the graph and the transitions between states are defined by considering the rates labeling the arcs. Since all activity durations are exponentially distributed, the total transition rate between two states will be the sum of the activity rates labeling arcs connecting the corresponding nodes in the DG. Starting from the DG of Figure 10.1, the derivation of the corresponding CTMC is straightforward and results in the generator matrix shown below.

$$Q = \begin{pmatrix} -\lambda & p_1\lambda & p_2\lambda & 0 & 0 \\ m & -m & 0 & 0 & 0 \\ 0 & 0 & -g & g & 0 \\ 0 & 0 & 0 & -\mu & \mu \\ r & 0 & 0 & 0 & -r \end{pmatrix}$$

In order to ensure that the underlying Markov process is ergodic, the DG of a PEPA model must be strongly connected. Necessary conditions for ergodicity, at the syntactic level of a PEPA model, have been defined in 14. For example, if cooperation occurs it must be the highest level combinator. The class of PEPA terms which satisfy these syntactic conditions are termed *cyclic components* and they can be described by the following grammar:

$$P ::= S \mid P \underset{L}{\bowtie} P \mid P/L$$
$$S ::= (\alpha, r).S \mid S + S \mid A$$

All the models we have discussed so far are cyclic models.

It is well known that if the CTMC is ergodic, it is possible to compute the steady state probability distribution over all the possible states by solving the matrix equation $\pi Q = 0$ where Q is the state transition rate matrix of the CTMC and π is the state probability vector, such that $\sum_i \pi_i = 1$.

The probability distribution of the states of the model is often not the ultimate goal of performance analysis. In order to define performance measures such as throughput and utilisation the modeler can define a reward structure over the CTMC at the level of the process algebra. This is done by associating rewards with actions: the reward associated with a state is then the total reward attached to the actions it enables. Note that no reward can be attached to internal, τ, actions.

We now describe in some detail a modified version of the previous example showing how we can extract performance measures from the model. We consider again a multiprocessor system with a shared memory in which all processes have the same functional behavior: they compete for access to the shared memory, use and then release it. In this case, however, we assume that basic actions progress at different speeds depending on the processor on which they are running, and that the action *think* models both the processing activity and the access to the local memory. Thus a process running on the ith processor has the following specification:

$$P_i \stackrel{\text{def}}{=} (think, \lambda_i).(get, g).(use, \mu_i).(rel, r).P_i$$

When n_i processes P_i are running on the ith processor the system is modeled by considering n_i independent replicas of the same process:

$$Sys_5 \stackrel{\text{def}}{=} (\underbrace{P_1 \| \cdots \| P_1}_{n_1} \| \cdots \| \underbrace{P_N \| \cdots \| P_N}_{n_N}) \bowtie_L Mem \qquad L = \{get, use, rel\}$$

The DG and the CTMC underlying the model can automatically be obtained using the PEPA Workbench[7] when we instantiate the model parameters, i.e. the number of processors and the number of processes. In the simple case of $N = 3$, $n_1 = n_2 = 2$, $n_3 = 1$ we obtain

$$Sys_6 \stackrel{\text{def}}{=} (P_1 \| P_1 \| P_2 \| P_2 \| P_3) \bowtie_L Mem \qquad L = \{get, use, rel\}$$

whose DG has 192 states.

The CTMC is derived directly from the DG by considering only the action rates labeling the arcs. By solving the set of linear equations we obtain the steady state probability vector. If we instantiate the rate parameters of the model as follows,

$$r = g = 100, \ \lambda_1 = 3, \ \lambda_2 = 2, \ \lambda_3 = 3, \ \mu_1 = 0.5 \times m, \ \mu_2 = \mu_3 = 3$$

where m takes values in the range $1, \ldots, 5$, and calculate, for each processor, the percentage of time it spends waiting for access to the memory relative to the time it spends accessing the memory, we obtain the graph shown in Figure 10.2.

The percentage waiting time for each processor is found by associating a reward of 1 with each derivative in which *get* is the only action enabled by any process on that processor, but the memory is not enabling *get*. The time accessing the memory is found by associating a reward 1 with each derivative in which any process is enabling *use* or *rel*, or enabling *get* when the memory is also enabling *get*. The graph suggests that the processes running on Processor 1 are monopolising the memory.

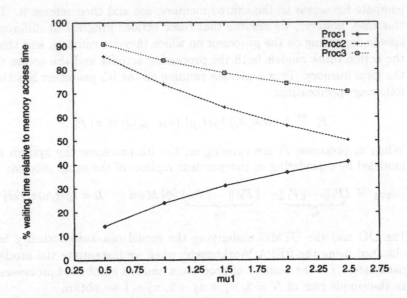

Figure 10.2: Percentage waiting times for Processors 1, 2 and 3 as the mean duration of Processor 1 memory access decreases

One way in which we might alleviate this problem would be to introduce a protocol for memory access which attempts to restrict the patterns of behavior exhibited by the processors. In the protocol, represented in the modified model shown below, after granting access to a given processor the memory will not grant access to that processor again until a different processor has been served. In the component representing the memory we now distinguish the *get* actions originating on different processors, in order to enforce the protocol. The memory also remembers which processor last had access in order to disallow it in the next access. Thus Mem_i represents the memory when it was last accessed by the ith processor and it is modeled as the sum of all its possible behaviors:

$$Mem_i \overset{\text{def}}{=} \sum_{\substack{j=1 \\ j \neq i}}^{N} (get_j, \top).(use, \mu_j).(rel, \top).Mem_j$$

Each process model is also modified by renaming its *get* action to get_i, if it is running on Processor i; in all other aspects it remains the same.

The model of a multiprocessor system with this memory access pro-

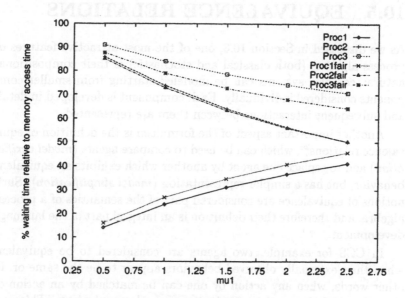

Figure 10.3: Percentage waiting times for Processors 1, 2 and 3 under both access protocols, as the mean duration of Processor 1 memory access decreases

tocol is given below.

$$Sys_5' \stackrel{\text{def}}{=} (\underbrace{P_1 \| \cdots \| P_1}_{n_1} \| \cdots \| \underbrace{P_N \| \cdots \| P_N}_{n_N}) \bowtie_L Mem_i \qquad L = \{get_i, use, rel\}$$

Note that with the inclusion of Mem_i in the initial system we are imposing that the starting state excludes access of an arbitrary processor $Proc_i$. When, as before, $N = 3$, $n_1 = n_2 = 2$, $n_3 = 1$ we have:

$$Sys_6' \stackrel{\text{def}}{=} (P_1 \| P_1 \| P_2 \| P_2 \| P_3) \bowtie_L Mem_1 \qquad L = \{get_1, get_2, get_3, rel, use\}$$

whose DG has 256 states. As previously, we calculate the percentage time each processor spends waiting for access relative to access time. The results, under both the original access protocol and the new fair protocol, are shown in Figure 10.3. As hoped, this modification to the system does reduce the percentage of time that Processors 2 and 3 spend waiting for access to the memory.

10.5   EQUIVALENCE RELATIONS

As we discussed in Section 10.3, one of the most attractive features of process algebras (both classical and stochastic) is their compositional nature: complex systems can be modeled starting from smaller components considered individually. Each component is developed in detail and subsequent interactions between them are represented.

Another important aspect of the formalism is the definition of equivalence relations[20], which can be used to compare agents (*model verification*) and to replace one agent by another which exhibits an equivalent behavior, but has a simpler representation (*model simplification*). Such notions of equivalence are considered part of the semantics of a process algebra, and therefore their definition is an integral part of the language development.

In CCS for example, two agents are considered to be equivalent when their externally observed behaviors appear to be the same or, in other words, when any action by one can be matched by an action of the other and the resulting derivatives are also equivalent. This is a formally defined notion of equivalence, based on the labeled transition system underlying the process algebra. Different equivalence relations have been proposed depending on whether the action τ is considered to be observable (*strong bisimulation*) or not (*weak bisimulation*)[20].

Such equivalence relations are particularly useful when they can be shown to be a *congruence* and therefore complementary to the structure of the model. A relation is a congruence if it is preserved by all the combinators of the language and this will mean that model verification and model simplification can be carried out on the components within a model, rather than across the whole model at once. If an algebraic characterisation (axiomatisation) is found, the axioms may be used to apply the equivalence relation at the level of the syntax rather than at the level of the underlying transition system.

By extending the standard theoretical results from classical to stochastic process algebras, different notions of equivalence have been defined. Here both functional and temporal aspects have to be taken into account, that is, both actions and delays have to be considered. In this context equivalence relations have been shown to have practical implications for performance modeling.

Currently in performance analysis notions of equivalence are used informally to address the problems of large Markov chains in two important ways. Firstly, an alternative model may be proposed; this could be easier to solve because it is more compact or exhibits special structure which is susceptible to efficient solution, e.g. product form. In this case a notion of equivalence *between models* may be used to establish whether or not the alternative model is an adequate representation of the original (*model simplification*). Secondly, equivalence *within a model* may identify states which exhibit sufficiently similar behavior that they need not be distinguished, thus reducing the size of the model to be solved (*aggregation*).

In queueing networks and stochastic Petri nets there has been little formal development of equivalence between models. In contrast there has been much work on the definition of equivalences at the state space level. These equivalences form the basis of aggregation techniques for reducing the state space of the underlying Markov chain, and thus provide a technique for making the solution of large models tractable.

Stochastic process algebras are the first performance modeling paradigm to offer a formalisation of both notions of equivalence. In the following subsections we discuss two distinct equivalence relations which have been developed for PEPA[14]. These can be used, respectively, for model simplification and aggregation.

10.5.1 Weak Isomorphism

The use of the hiding operator during the development of a PEPA model allows the modeler to represent aspects of a component's behavior in detail but to later abstract away from it. Such restriction may also be used to safeguard the interaction between components from external interference, as we saw in Section 10.3. This style of modeling may mean that, when complete, a model executes sequences of τ, or internal, actions. Since rewards cannot be attached to such actions no performance information is obtained from these sequences. It is natural therefore to try to simplify them.

The most straightforward means of simplifying a sequence of hidden actions is to replace them by a single hidden action, thus amalgamating intermediate states. Such an approach would, in general, lead to an approximation technique, and has not yet been developed. However an equivalence relation called *weak isomorphism* has been developed for PEPA which can be used to identify when such an approach will achieve exact model simplification. Using weak isomorphism the observation of internal activities is relaxed, and insensitive sequences of τ actions may be replaced by a single τ action with appropriate mean.

The calculation of different performance measures may require different action names to be visible or hidden. We can think of this as defining an experimental context for the model. Via judicious use of the PEPA abstraction mechanisms, weak isomorphism allows a model to be modified to a simpler form, reflecting its current experimental context. Although the weak isomorphism relation is not a congruence relation, it has been shown to be preserved by cooperation, and so this model simplification technique can be applied compositionally in cyclic models.

10.5.2 Strong Equivalence

Strong equivalence is a bisimulation which aims to capture when two PEPA components are indistinguishable under experimentation. This equivalence notion considers both the functional and the temporal aspects of the model but observation occurs without detailed knowledge of the individual actions involved. Thus strong equivalence is unable to distinguish between a single $(\alpha, 2r)$ activity and two simultaneously enabled (α, r) activities.

The concept of *conditional transition rate*, $q(C_i, C_j, \alpha)$, is the basis for the definition of strong equivalence. This is the rate at which a system behaving as component C_i evolves to behaving as component C_j as a result of completing an activity of type α. If we consider a *set* of possible derivatives S we can compute the *total conditional transition rate* from C_i to S as the sum of the conditional transition rates from C_i to any component C_j in the set S

$$q[C_i, S, \alpha] = \sum_{C_j \in S} q(C_i, C_j, \alpha)$$

Two PEPA components are considered strongly equivalent if there is an equivalence relation between them such that, for any action type α, the total conditional transition rates from those components to any equivalence class, via activities of this type, are the same.

The formal definition of strong equivalence may be found in 14 where it has been shown that this relation is sufficient to ensure that the components exhibit the same behavior. Moreover, strong equivalence may also be used over the state space underlying a single PEPA model, resulting in a lumpable partition of the state space. The DG of the model is partitioned into equivalence classes and each equivalence class forms a single state in the aggregated Markov process.

Thus the well-known technique of aggregation based on lumpability is now defined at the level of the modeling paradigm rather than of the underlying CTMC. However this implies that the entire state space must be generated. This *a priori* generation of the state space is undesirable and can be avoided because the strong equivalence relation is a congruence.

A procedure for implementing strong equivalence aggregation during the composition of cooperating components is outlined in 16. Partial state spaces (corresponding to cooperating components) are generated and aggregated. Then each pair of cooperating components is replaced by a corresponding aggregated model and the procedure is repeated. Thus the state space of the complete model does not need to be generated. Moreover the formality of the approach has important implications for tool support for model simplification.

We now show how strong equivalence may be used to reduce the state space of the Markov process underlying a PEPA model. We consider again the multiprocessor system with the memory access protocol we discussed in Section 10.4:

$$Sys_6' \stackrel{\text{def}}{=} (P_1 \| P_1 \| P_2 \| P_2 \| P_3) \underset{L}{\bowtie} Mem_1 \qquad L = \{get_1, get_2, get_3, rel, use\}$$

Recall that the DG of Sys_6' has 256 states. By applying strong equivalence we can recognise that some states belong to the same equivalence class. For example, if one of the processes on Processor 1 is thinking and the other is using the memory, the behavior of the system is the same as if their roles were reversed, i.e. the first is using the memory and the other is thinking. Exploiting such symmetries between processes running on the same processor we can reduce the state space to 120 states. Note that this is an exact aggregation and the performance measures extracted from the model are unaffected. This corresponds to lumping together states which differ only in a permutation of the derivatives of the components modeling processes running on the same processor. This is shown in diagrammatic form in Figure 10.4, for the case of the two processes P_1. The two states numbered #1 and #2 are strongly equivalent and therefore they belong to the same equivalence class. Similarly for the two states #3 and #4.

Notice that the reductions we have obtained on the example are based strictly on the symmetries of the model which occur because of

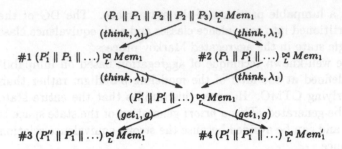

Figure 10.4: Small portion of the derivation graph of Sys_6'.

repeated components. The power of strong equivalence is not limited
to this case. For example, consider a model in which the get_i actions
are hidden and considered internal to the model. If it is also the case
that processes running on the different processors share the same characteristics, i.e. $\lambda_1 = \lambda_2 = \lambda_3$ and $\mu_1 = \mu_2 = \mu_3$, then we are able
to obtain a further reduction by lumping more states. This further
reduction can be explained by the observation that only one process
accesses the memory at a time and it does not really matter which:
other processes are indistinguishable. Once the get_i actions become τ
actions, and therefore indistinguishable, strong equivalence will result
in a further substantial reduction of the state space, to 67 states.

In general, strong equivalence can also be used to find more general
forms of repeated patterns of behavior within models. When repeated
patterns of behavior are recognised, the states that have the same behavior (i.e. states that after performing the same actions with the same
rates reach equivalent states) are lumped together.

10.6   OTHER SPA

Work on SPA originated with Herzog's group at the University of Erlangen. The initial work was carried out on a process algebra EXL
which was a variant of CSP in which a random variable was associated with each event and a probabilistic choice operator replaces nondeterministic choice[13]. This language later evolved into TIPP (TImed
Process for Performance Evaluation)[8,9]. The work on TIPP has been
primarily motivated by a desire to encourage the timely consideration
of performance characteristics during system development, particularly
for distributed systems[13]. Although originally TIPP was studied un-

der very general assumptions that allowed any distribution to be used to characterise the duration of an action, difficulties in extracting performance measures from such models has meant that more recent work has focussed on a Markovian version of the language (MTIPP)[11] which is very close to PEPA.

Markovian Process Algebra (MPA) was developed by Buchholz at the University of Dortmund[4,5] in 1994. The major difference between this language and PEPA is the assumption that all actions of type α proceed at a fixed rate μ_α. Activities are still represented as (α, r) pairs but r now represents the number of concurrently enabled instances of action α, all of which proceed at rate μ_α. This difference has a subtle impact on the semantics of the language.

Extended Markovian Process Algebra (EMPA) was also developed in 1994 by Gorrieri's group at the University of Bologna[1,2]. Originally this SPA was also called MPA but the name was later changed to avoid ambiguity. The language includes a rich set of combinators. It was the first SPA to include *immediate* or instantaneous actions. Passive actions play a central role in the theory of EMPA and synchronisation is restricted to involve at most one timed or immediate action. Most of the work on EMPA has focussed on Petri net based semantics for the language[3].

The most significant difference between the SPA languages lies in their different definitions of synchronisation. This is discussed in some detail in 15.

10.7 EFFICIENT SOLUTION

SPA models are prone to the problem of state space explosion: even moderately sized models can generate a huge state space making analysis of the system infeasible. Much of the recent work, like the earlier work on equivalence relations, is motivated by a desire to address this problem. Two main strands are identifiable in the current work. The first seeks to characterise models in which the compositional structure may be exploited in order to gain efficient solution of the underlying CTMC by considering submodels in isolation. The second aims to avoid the construction of large CTMCs by removing the condition that each syntactic term encountered in the model is treated as a distinct state.

Given the clear component-based structure of SPA models it is perhaps natural to consider decompositional solution techniques. A class of SPA models susceptible to such a solution technique is presented in 17. Syntactic conditions to identify models which satisfy Courtois's *near complete decomposability* property[6] are presented, meaning that time scale decomposition can be carried out automatically without the need to ever construct and store the whole state space.

Efficient product form solution is one of the major attractions of queueing networks for performance modeling purposes. These models rely on a form of interaction between nodes in a network which allows them to be solved in isolation, since they behave as if independent, up to normalisation. In a recent paper[10], the results from queueing networks have been exploited to identify a restricted form of interaction between suitable SPA components which leads to a product form solution. Another approach, based on work on stochastic Petri nets which satisfy certain structural conditions, has identified another class of product form SPA models[25]. In such models the actions themselves may be considered to be states of a Markov chain which may be solved to find the steady state of the complete process.

Incorporating immediate actions into a model means that "states" enabling such actions can be eliminated. In 12, MTIPP is enriched with immediate actions which allow control activities to be modeled as actions which take a negligible amount of time. In order to derive the underlying CTMC, immediate actions must be removed. This procedure, based on a relation called Markovian observational congruence[12], is only possible for immediate actions which are internal, i.e. ones which do not participate in external synchronisation because they are hidden from the environment. The congruence property ensures that the elimination can be performed on the fly, during the construction of the CTMC.

Immediate actions can also be used to model branching probabilities (as in Generalized Stochastic Petri Nets). In 23 this use of immediate actions is proposed for MTIPP, via the introduction of a distinct operator to represent probabilistic branching. Again, a specific congruence relation is the key to correct derivation of the underlying CTMC.

An alternative compositional approach to the solution of Markov processes is the use of tensor algebra to express the generator matrix of a process[22]. This approach is based on a restricted form of interaction between subsystems within the system and it has also been applied to SPA[4,24].

10.8 CONCLUSIONS

In this chapter we have described an algebraic description technique, based on a classical process algebra, and enhanced with timing information. This extension results in models which may be used to calculate performance measures as well as deduce functional properties of the system. Several interesting features of SPA, such as the compositionality, the abstraction mechanism, and the equivalence notions suitable for performance analysis, have been discussed in some detail. This approach to performance modeling is a recent innovation and a lot of research activity is presently going on as we have indicated in Section 10.7. At the same time tool prototypes are under development to exploit the possibilities for automation presented by the SPA approach.

REFERENCES

1. M. BERNARDO, L. DONATIELLO, and R. GORRIERI, "Modeling and Analyzing Concurrent Systems with MPA," Proc. 2nd Process Algebra and Performance Modeling Workshop, (Erlangen, July 1994).

2. M. BERNARDO, L. DONATIELLO, and R. GORRIERI, "MPA: a Stochastic Process Algebra," Technical Report UBLCS-94-10, University of Bologna, Laboratory of Computer Science, (May 1994).

3. M. BERNARDO, L. DONATIELLO, and R. GORRIERI, "Giving a Net Semantics to Markovian Process Algebras," Proc. 6th Int. Workshop on Petri Nets and Performance Models, (Durham, NC, October 1995).
   The papers 1,2, and 3 discuss the EMPA language. The first two describe the language; the third provides both an operational and a denotational net semantics for it.

4. P. BUCHHOLZ, "Markovian Process Algebra: Composition and Equivalence," Proc. 2nd Process Algebra and Performance Modeling Workshop, (Erlangen, July 1994).

5. P. BUCHHOLZ, "On a Markovian Process Algebra," Technical Report 500/1994, University of Dortmund, (1994).
   The two papers 5 and 6 describe the MPA language with particular emphasis on the compositional generation of the underlying Markov process using tensor algebra.

6. P.J. COURTOIS, "Decomposability: Queueing and Computer Sys-

tem Applications," ACM Series, Academic Press, (New York, 1977).

7. S. GILMORE and J. HILLSTON, "The PEPA workbench: A Tool to Support a Process Algebra Based Approach to Performance Modeling," Proc. 7th Int. Conf. on Modeling Techniques and Tools for Computer Performance Evaluation, (Vienna, 1994).

This paper describes the PEPA workbench, a prototype which provides a set of simple tools to delegate to machine assistance some of the routine tasks in checking PEPA descriptions, generating the underlying CTMC and calculating performance measures.

8. N. GOTZ, U. HERZOG, and M. RETTELBACH, "TIPP - Introduction and Application to Protocol Performance Analysis," Technical Report, University of Erlangen, (March 1993).

9. N. GOTZ, U. HERZOG, and M. RETTELBACH, "Multiprocessor and Distributed System Design: The Integration of Functional Specification and Performance Analysis Using Stochastic Process Algebra," Tutorial Proc. Performance 1993, vol. 729 of LNCS, (Rome, 1994).

The two papers above present the TIPP language. In 9 general distribution functions are allowed and the difficulty of extracting performance measures in this case are discussed. In 10 the distribution of the delays is restricted to be the negative exponential distribution function.

10. P.G. HARRISON and J. HILLSTON, "Exploiting Quasi-reversible Structures in Markovian Process Algebra Models," The Computer Journal, 38(7), 1995. Special Issue: Proc. 3rd Process Algebra and Performance Modeling Workshop, (Edinburgh, June 1995).

11. H. HERMANNS and M. RETTELBACH, "Syntax, Semantics, Equivalences, and Axioms for MTIPP," Proc. 2nd Process Algebra and Performance Modeling Workshop, (Erlangen, July 1994).

This paper presents a notion of equivalence, called Markovian bisimulation, which is essentially the same as strong equivalence. In addition an equational theory for this equivalence is presented for models with finite (i.e. terminating) behavior. This technique allows model aggregation via syntactic manipulation of the process algebra description (term rewriting).

12. H. HERMANNS, M. RETTELBACH, and T. WEISS, "Formal Characterisation of Immediate Actions in SPA with Nondeterministic Branching," The Computer Journal, 38(7), 1995. Special Issue: Proc.

3rd Process Algebra and Performance Modeling Workshop, (Edinburgh, June 1995).

13. U. HERZOG, "Formal Description, Time and Performance Analysis: A framework," Technical Report 15/90, IMMD VII, Friedrich-Alexander-Universität, Erlangen-Nürnberg, (Germany, September 1990).

14. J. HILLSTON, "A Compositional Approach to Performance Modeling," PhD thesis CST-107-94, Computer Science Department, University of Edinburgh, (April 1994).

All the information on the PEPA language, the equivalence relations, and the computation of performance measures may be found in this PhD thesis which will be published by the Cambridge University Press in 1996.

15. J. HILLSTON, "The Nature of Synchronisation," Proc. 2nd Process Algebra and Performance Modeling Workshop, (Erlangen, July 1994).

This paper compares the different SPA languages with particular emphasis on the various combinators used to model synchronisation between components.

16. J. HILLSTON, "Compositional Markovian Modeling Using a Process Algebra," Proc. 2nd Int. Workshop on the Numerical Solution of Markov Chains, (Raleigh, NC, January 1995).

This paper describes a procedure for the compositional derivation of an aggregated Markov process underlying a PEPA model that avoids the generation of the whole state space.

17. J. HILLSTON and V. MERTSIOTAKIS, "A Simple Time Scale Decomposition Technique for Stochastic Process Algebras," The Computer Journal, 38(7), 1995. Special Issue: Proc. 3rd Process Algebra and Performance Modeling Workshop, (Edinburgh, June 1995).

18. C.A.R. HOARE, "Communicating Sequential Processes," Prentice-Hall, (1985).

19. C-C. JOU and S.A. SMOLKA, "Equivalences, Congruences and Complete Axiomatizations of Probabilistic Processes," CONCUR'90, vol. 458 of LNCS, pp. 367–383. Springer-Verlag, (August 1990).

20. R. MILNER, "Communication and Concurrency," Prentice Hall, 1989.

This book describes CCS and the main features of the language. It is a good starting point for anyone wishing to learn about process algebra.

21. X. NICOLLIN and J. SIFAKIS, "An Overview and Synthesis on Timed Process Algebra," Real-Time: Theory in Practice, vol. 600 of LNCS, Springer Verlag, (1991).
This paper compares different timed extensions of process algebra. These extensions differ from SPAs since time progression and action execution are orthogonal concepts.

22. B. PLATEAU, J-M. FOURNEAU, and K-H. LEE, "PEPS: A Package for Solving Complex Markov Models of Parallel Systems," Proc. 4th Int. Conf. on Modeling Techniques and Tools for Computer Performance Evaluation, pp. 341–360, (1988).

23. M. RETTELBACH, "Probabilistic Branching in Markovian Process Algebras," The Computer Journal, 38(7), 1995. Special Issue: Proc. 3rd Process Algebra and Performance Modeling Workshop, (Edinburgh, June 1995).

24. M.L. RETTELBACH and M. SIEGLE, "Compositional Minimal Semantics for the Stochastic Process Algebra TIPP," Proc. 2nd Process Algebra and Performance Modeling Workshop, (Erlangen, July 1994).

25. M. SERENO, "Towards a Product Form Solution for Stochastic Process Algebras," The Computer Journal, 38(7), 1995. Special Issue: Proc. 3rd Process Algebra and Performance Modeling Workshop, (Edinburgh, June 1995).

CHAPTER 11

PERFORMANCE EVALUATION USING MICRO-BENCHMARKING AND MACHINE ANALYSIS

Rafael H. Saavedra

Alan Jay Smith

11.1. SUMMARY

We present a new methodology for CPU performance evaluation based on the concept of an abstract machine model and contrast it with benchmarking. The model consists of a set of abstract parameters representing the basic operations and constructs supported by a particular programming language. The model is machine-independent, and is thus a convenient medium for comparing machines with different instruction sets. A special program, called the machine characterizer, is used to measure the execution times of all abstract parameters. Frequency counts of execution parameters are obtained by instrumenting and running programs of interest. By combining the machine and program characterizations we can and do obtain accurate execution time predictions. This abstract model also permits us to formalize concepts like machine and program similarity.

A wide variety of computers, from low-end workstations to high-end supercomputers, have been analyzed, as have a large number of standard benchmark programs, including the SPEC scientific benchmarks. We present many of these results, and use them to discuss variations in machine performance and weaknesses in individual benchmarks. We explain how the basic model can be extended to account for the effects of compiler optimization, memory hierarchy, and vectorization.

11.2. INTRODUCTION

Comparing the CPU performance of different machines is a problem that has confronted designers and users for many years. Nowadays, the most widely used method is benchmarking. It consists of running a set of programs and measuring their execution times[15,11]. The advantage to benchmarking is that it yields measurements of real programs running on real computers. There are several shortcomings to benchmarking, however, one of which has been the problem that many standard benchmarks, e.g. Dhrystone[28], are considered to be substantially unrepresentative of 'normal' workloads. Three serious efforts to create a set of realistic benchmarks are the SPEC[24,26], Perfect Club[8], and the SPLASH[23,29] suites. However, even when a set of benchmarks is carefully assembled, there are some limitations to the technique[30,9]: 1) It is very difficult to explain benchmark results from the characteristics of the machines. 2) It is not clear how to combine individual measurements to obtain a meaningful evaluation of various systems. 3) Using benchmark results it is not possible to predict and/or extrapolate the expected performance for arbitrary programs. 4) The benchmarks may still not be representative, and/or may not do the kind of computation (e.g. integer vs. floating point) expected. 5) The large variability in the performance of highly optimized computers is difficult to characterize with benchmarks.

Another approach to machine performance evaluation is to model the machine at the instruction set level. This approach suffers from several problems: (a) It is difficult to construct an accurate model of machine operation, including all pipeline delays. (b) A very large number of parameters are needed for such a model. (c) The frequency of machine operations may be hard to know or estimate. (d) The model is different for every machine architecture, and in fact for every different implementation of each machine architecture.

We can consider benchmarking and machine models as being located at opposite extremes of a spectrum. Machine models are limited by their complexity and machine specificity. One the other hand, benchmarking lacks any kind of model, and without it, it is not possible to predict or explain benchmark results.

Our particular approach and the subject of this chapter has been to develop a new methodology which attempts to overcome the limitations of both benchmarking and machine models, while retaining their particular advantages. Our solution, which we call *Abstract Machine Performance Characterization*, consists of building a machine-independent model based on the set of operations used in source programs. Because the model is machine-independent, it applies to all computer systems running that programming language, independent of their particular instruction sets.

Machine characterization is accomplished by measuring the execution time of individual parameters via 'narrow spectrum' benchmarking. This produces a vector of measurements for all the abstract operations that determine the execution time of programs. We characterize programs by the number and type of abstract operations they execute. Using the set of measurements from the machines and programs, both expressed in terms of the same abstract model, it is simple to combine these to produce execution time estimates for arbitrary machine-program combinations. Execution time can also be decomposed in terms of what the programs does and how the machine performs with respect to individual operations. This relation between machine characterization, program characterization, and execution time can be expressed with following equation.

$$T_{A,M} = \sum_{i=1}^{n} C_{A,i} P_{M,i} = \mathbf{C_A} \cdot \mathbf{P_M} \qquad (11.1)$$

where $\mathbf{P_M} = <P_1, P_2, \cdots, P_n>$ represents the machine characterization, $\mathbf{C_A} = <C_1, C_2, \cdots, C_n>$ represents the program characterization, and $T_{A,M}$ the execution time of program A on machine M. A graphical representation of this process can be seen in Fig. 11.1.

It is important to stress that we don't consider our approach and benchmarking to be incompatible. On the contrary, we regard our approach as a more general methodology encompassing benchmarking and building on it. The abstract machine model has several advantages not present in benchmarking or machine models. Some of these are: 1) A single benchmark (the machine characterizer) is used to completely characterize the performance of a machine, instead of using a large set of benchmarks that only provide unrelated observations of performance. 2) It is possible to compare different architectures or implementation of the same architecture in a machine independent way, and to study how different machines react to the most time-consuming sections of the program. 3) Machines can be compared at many different levels: individual abstract operations, functional units, programs, and workloads. 4) The machine abstract model can be used to easily select, from a large set of machines, those which satisfy some performance requirements. 5) Making a detailed analysis of benchmarks helps to understand those machine features that they test. 6) Benchmarking real machines is a time consuming activity; running N benchmarks in M machines require $N \cdot M$ steps. In contrast using the abstract machine model requires only $N + M$ machine and program characterizations[†]. 7) We are able to specify proposed or unimplemented machines (by choosing appropriate values for the abstract parameters), and/or specify proposed or anticipated workloads (by choosing appropriate frequencies for the abstract parameters). In this way, we

---

[†] Although we still need to produce $N \cdot M$ predictions, the amount of work require to do this is negligible.

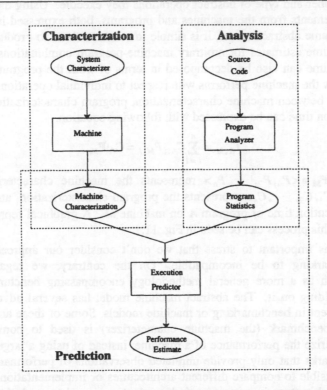

Figure 11.1: The process of characterization, analysis and prediction. There exists one performance vector (P_M) for each machine, and a set of statistics (C_A for each program. Execution time estimates are obtained by merging the machine characterizations with the program statistics.

can estimate the performance of arbitrary real or unreal machines on existing or future workloads.

In what follows we discuss the abstract machine model in more detail, showing how it is used to characterize and compare machines, programs, and optimizing compilers. We give measurements for real machines and programs, and discuss the significance of some of the results. Then we present metrics for program and machine similarity. These allow us to quantify the extent to which performance on two machines will be proportional across a range of programs. We finish by presenting some of the work we are doing in characterizing optimizing compilers. We note that this chapter is only a short summary overview of our work on this topic; further information can be found in refs. 18-22. Related work on parallel and vector machines appears in ref. 27.

11.3. THE EFFECTS OF THE MEMORY HIERARCHY, COMPILER OPTIMIZATION, AND VECTORIZATION

Most of the discussion in this chapter is based on our basic model which ignores the effects of compiler optimization, memory hierarchy, and vectorization. For completeness purposes, however, in this section we briefly explain how the model can be extended to account for these effects. We also indicate, when appropriately, how these factors affect our ability to predict the execution time of programs. More details are given in the appropriate references.

11.3.1. Memory Hierarchy

To account for the effects of the memory hierarchy we can add an explicit term to (11.1) and get

$$T_{A,M} = \sum_{i=1}^{n} C_{A,i} P_{M,i} + \sum_{i=1}^{m} M_{A,M,i} S_{M,i} = C_A \cdot P_M + M_{A,M} \cdot S_M, (11.2)$$

where $M_{A,M}$ and S_M represent vectors of misses and stall time penalties affecting all relevant levels of the memory hierarchy[22][†].

In contrast with (11.1) where we use a single program characterization on all machines, accounting for the memory delays requires that each program and memory structure combination be characterized with both miss ratios and delay times for each level or component of the memory hierarchy.

In this chapter we ignore the effects of the memory hierarchy, but doing this does not significantly affect our predictions. The reason for this is that almost all the SPEC and Perfect benchmarks suffer from low amounts of cache, TLB, and page misses/faults. In ref. 22 we show that, on the SPEC benchmarks, including the effects of the cache and TLB misses account, on the average, for no more than 7.09% and 1.03%, respectively, of the total execution time. (Miss ratios for only a few of the benchmark programs that we study are available[10].) With respect to cache misses, the maximum delay contribution occurs on *NASA7* with 8.55%, and with respect to TLB misses, the maximum contribution is for *MATRIX300* with 3.09%. Overall, the effect of the memory hierarchy on

---

[†] The TLB is considered also a level in the hierarchy, even when in reality it is not.

our prediction errors reduces the root mean square error by 2.36%. Therefore, ignoring memory hierarchy effects does not prevent us from validat-
ing the predictions of the simpler model.

11.3.2. Optimization

In this chapter we consider only non-optimized execution in both our
predictions and run-time measurements. When dealing with optimization,
the problem for any machine-independent model is to "guess" how an
arbitrary optimizing compiler will change the source code without having
to know the internal organization of the compiler. Note that the need to
guess the actions of the optimizer is not a requirement of our model, but a
limitation imposed by the fact that compilers do not provide information
about how the the optimization process affects programs. If this information were available, our model could easily use it to adjust its predictions.

At the same time it is clear that any practical model for program execution has to be able to predict the execution time of optimized code.
Therefore, we have dealt with this problem explicitly in an extension of
this work[19,21]. In general terms, we can say that compilers attempt to
optimize the code each time the program is translated from a higher to
lower representation. In this respect, the optimizations attempted by most
compiler can be classified as: high-level optimizations[†] (source language
dependent), low-level optimizations (source and target language independent), and micro optimizations (target language dependent)[2]. The high-
level optimization opportunities arise from the algorithmic characteristics
of the program, and although some transformations can drastically alter
the program look, the resulting code continues to represent a valid
source-level program. As such, the optimized version can be characterized, in terms of our abstract machine model, in the same way as the original unoptimized program producing an optimized program characterization $C_{A,O}$[‡].

Low-level and micro optimizations cannot be handled in the same
way as high-level optimizations, because the effects of these optimizations
cannot be represented by an equivalent source-level program. Their
effects can still be quantified by considering that the aggregate effects of
low-level optimizations can be considered as defining an "optimized"

[†] Based on this classification optimizing pre-processors also apply high-level optimizations.

[‡] Although in principle this is true, in practice this may require changing the compiler to make it possible to see the optimized program in the original high-level
language. Most optimizing pre-processors do this by default.

machine running the same sequence of abstract machine instructions. The performance of this optimized abstract machine is characterized by a new performance vector $P_{M,O}$. This approach works because low-level optimizations tend to improve sequence of low-level instructions implementing high-level ones, without eliminating the semantics of the original instructions. In ref. 21 formalize this notion by introducing the concept of *invariant optimization*.

Once the effects of high-level, low-level, and micro optimizations have been captured with vectors $C_{A,O}$ and $P_{M,O}$, the execution time of an optimized program can be predicted using (11.1). In ref. 21 we show that the rms prediction errors for nonoptimized and optimized programs are 20.59% and 31.89%, respectively. The error increase is quite small considering that the amount of improvement due to optimization range from less to 40% to as much as 80%. More details are given in refs. 19, 21.

11.3.3. Vectorization

Extending our model to include vector machines is straightforward and consists of two steps. First, a set of vector AbOps is added to the original set which characterizes the performance of individual vector operations. In contrast with scalar AbOps, which are characterized with a single number, vector AbOps require three numbers: the startup delay, the maximum vector length, and the asymptotic latency. Second, vectorizing a program is similar to applying a high-level optimization, in which some statements inside the innermost loops are transformed into a sequence vector statements[†]. The resulting vectorized program represents a valid Fortran program written in vector notation. Hence, computing the dynamic statistics for a vectorized program is not more difficult than doing it for a scalar one. Finally, an equation similar to (11.1) can be used to make execution time predictions. Some work along these lines is ref. 27.

11.4. THE ABSTRACT MACHINE MODEL

Every programming language can be viewed as defining an abstract machine model, as specified in its language constructs and basic operations. Therefore, a programming language allows us to consider different machines as emulators of a single abstract machine. Execution time

---

[†] If the length of the loop is larger than the vector length, extra code for strip mining the original loop is also added.

depends only on the compiler and the underlying machine and thus programming languages provide an ideal vehicle for building machine-independent models.

In our research, we have focussed on performance in a scientific environment, and have therefore built our model around the Fortran language; despite its antiquity, Fortran is still the most widely used programming language for scientific applications, and the majority of standard benchmarks are in Fortran. A similar abstract model could be created for most other algorithmic languages. In order to distinguish as much as possible between the compiler and the underlying machine, we have initially considered only unoptimized code; at the end of this chapter, we discuss and evaluate optimizing compilers.

Our Fortran abstract machine model contains 109 parameters. Each can be classified in one of the following broad categories: arithmetic and logical, procedure calls, array references, branching and iteration, and intrinsic functions. For example, there are individual parameters for each of the arithmetic operations (add/subtract, multiply, division, and exponentiation), for each data type (real, integer, and complex), and in most cases for each precision (single, double). We decided which parameters to include in our model in an iterative manner. Initially we associated parameters with obvious basic operations, and after a first version of the system was running, new parameters were incorporated as needed to distinguish between different uses and execution times of the 'same' abstract operation in the program. Thus, the number of parameters has increased from 74^{17}, to 102^{19}, and to 109 currently. Although every basic operation in Fortran is characterized by some parameter, we have made simplifications in operations which were rarely executed in the benchmarks we used. It is important to note that the 109 parameters are obtained by direct measurement, and not by any sort of curve fitting; we have not been simply increasing the number of parameters in a fitted model.

We can illustrate how the abstract parameters of the model relate to characteristics of the machines and programs with an example. The following code fragment represents one of the most executed basic blocks found in the NAS Kernels[3].

```
      DO 1 I = 1, L
        C(I,K) = C(I,K) + A(I,J)*B(J,K) + A(I,J+1)*B(J+1,K) +
  1               A(I,J+2)*B(J+2,K) + A(I,J+3)*B(J+3,K)
  1   CONTINUE
```

Static analysis decomposes the statements into abstract operations and associates these with a unique basic block. Hence the code can be 'compiled' into the set of abstract operations shown in Table 11.1.

Opers.	Mnem	Description
1	LOIN	DO LOOP initialization and bounds check
1	SRDL	store of a double precision real
4	ARDL	double precision floating point add
4	MRDL	double precision floating point multiply
10	ARR2	references to 2-D array elements
6	ADDI	add between an index and a constant
1	LOOV	DO LOOP overhead: increment, check, branch

Table 11.1: Static statistics for one of the loops in the NAS Kernels.

By supplementing the static analysis given above with dynamic counters indicating the number of times each basic block is executed, we obtain a count of the number of times each operation occurs. Given the time for each operation, we can then compute the execution time of this source code fragment.

$$Time = N_1 \cdot T_{LOIN} + N_2 (T_{SRDL} + 4 \cdot T_{ARDL} + 4 \cdot T_{MRDL} + 10 \cdot T_{ARR2} + 6 \cdot T_{ADDI} + T_{LOOV}),$$

where N_1 corresponds to the number of times that the basic block containing the loop executes, and N_2 is the number of times that the body of the loop executes.

11.5. MACHINE CHARACTERIZER

The machine characterizer (MC) consists of 109 'software experiments' that measure the performance of each individual abstract parameter. The MC is written as a Fortran program and runs from 200 seconds, on machines with good clock resolution, to 2000 seconds on machines with 1/60'th or /100'th second resolution. We have run the MC on many different machines ranging from low-end workstations to supercomputers. Each experiment tries to measure the execution time that each parameter takes to execute in 'typical' Fortran programs. This typical execution time was obtained by looking at real programs and also by modifying those experiments that were identified as generating the biggest error in our predictions.

The general approach to measuring the execution times of parameters has been to time two versions of a loop, one with the parameter of interest in it, and one without. The main difficulties in measuring the performance of abstract parameters lie in their very small execution times, ranging from nanoseconds to hundreds of microseconds and the crudeness of the timing tools available in most machines; clocks normally have a resolution of only 1/60th or 1/100th of a second. Accuracy is obtained by running through the loop many times, and then running the overall loop test itself many times. Some parameters can't be measured quite so easily, and must be derived as the difference of other operation times[19]; for example, 'load' is not a Fortran operation. Despite a careful statistical approach to these measurements, there are some residual sources of error: the resolution, overhead and intrusiveness of the measuring tools; external events like interrupts, multiprogramming and I/O activity; variations in the hit ratio of the memory cache, and paging[6].

11.6. FROM BASIC TO REDUCED PARAMETERS

Vector P_M in (11.1) corresponds to the characterization of machine M in terms of the Fortran basic operations. It represents our fundamental measurement of performance, and from it, all predictions and metrics are computed. Unfortunately, it is very difficult to understand the meaning of an 109 element vector. For this reason we have defined a set of seventeen 'reduced' parameters that consolidate the original 109. The reduced parameters are obtained from the basic ones by aggregating those that exercise a similar functional unit in the processor and assigning them weights according to how frequently they are executed by programs.

Reduced Parameters

1 memory bandwidth (single)	10 double precision arithmetic
2 memory bandwidth (double)	11 intrinsic functions (single)
3 integer addition	12 intrinsic functions (double)
4 integer multiplication	13 logical operations
5 integer arithmetic	14 pipelining
6 floating point addition	15 procedure calls
7 floating point multiplication	16 address computation
8 floating point arithmetic	17 iteration
9 complex arithmetic	

Table 11.2: The seventeen reduced parameters. Integer and floating point arithmetic refer to all arithmetic operations, except addition and multiplication.

Table 11.2 shows the list of the seventeen reduced parameters. Most of the parameters deal with arithmetic characteristics, as would be expected for a language like Fortran. There are hardware, software, and hybrid parameters. Hybrid parameters are those that are implemented in hardware on some machine and in software on others. Parameters characterizing hardware functional units are: integer addition and multiplication, logical operations, procedure calls, looping, pipelining, and memory bandwidth (single and double precision). Pipelining comprises the different types of Fortran 'goto' statements. Software characteristics are represented by (intrinsic) trigonometric functions (single and double precision). Floating point, double precision and complex arithmetic, and address computation belong to the hybrid class. Note that these 17 combined parameters are not orthogonal, and in some cases more than one reduced operation is performed by the same functional unit.

A very convenient graphical way of representing the reduced parameters is to use a modified version of Kiviat graphs. Here the absolute or normalized values of the parameters are plotted on a logarithmic scale around a circle. In this way we can convey very effectively how different machines distribute their performance. We call these figures *pershapes* (performance shapes), and each system has a unique pershape. Figure 11.2 shows the pershapes for 2 supercomputers, 2 mainframes, 11 workstations and several implementations of the VAX architecture. The performance differences between two compilers or the impact a floating point co-processor are clearly reflected in the different pershapes. For example, in Fig. 11.2 we see how the use of two different Fortran compilers determine the pershape, as in the case of the *fort* and *f77* compiler on the VAX-11/785. In a similar way, the impact of using a floating point co-processor is clearly seen on the pershapes for the Sun 3/260 (the one with the co-processor is indicated by an (f)).

We can trace the evolution of performance in workstations by looking at the pershapes. If we compare the pershapes for the VAX-11/780 and the Sun 3/50, we see that for floating point arithmetic (single and double precision), complex arithmetic, intrinsic functions, and logical operations the Sun 3/50 had worse performance than the VAX-11/780. Even in the Sun 3/260 with a co-processor (SUN 3/260 (f) in the figure), single precision and complex arithmetic lagged behind that of the VAX-11/780. However, newer workstations, such as the IBM RS/6000 530, show almost two orders of magnitude improvement with respect to the Sun 3/260 in floating point and complex arithmetic. In contrast the performance improvement in integer and similar operations has been only one order of magnitude.

Pershapes can be used to compare the relative performance of machines, and determine their similarity. The idea is that two similar machines A and B will execute any arbitrary program P, K times faster on A than B. For dissimilar machines, for one program A may be faster, and for another, B may be faster. We define machine similarity (or pershape similarity) as the distance between two different performance shapes. We have created a similarity metric, shown below in (3), which is based on the 17 reduced parameters, and that has the right properties[18].

Figure 11.2: Performance of the reduced parameters with respect to the VAX-11/780. The concentric circles represent .1, 1, 10, and 100 times faster. The closest a performance shape (pershape) is to a circle, the closest the machine is to a VAX-11/780 in terms of how both machines distributed their performance along different computational modes.

$$d(X,Y) = \left[\frac{1}{n-1} \sum_{i=1}^{n} \left(\log(\frac{x_i}{y_i}) - \frac{1}{n} \sum_{j=1}^{n} \log(\frac{x_j}{y_j}) \right)^2 \right]^{1/2}. \qquad (11.3)$$

Vectors $X = <x_1, x_2, \cdots, x_n>$ and $Y = <y_1, y_2, \cdots, y_n>$ are two reduced performance vectors in $(0, \infty)^n$ representing the pershapes of machines M_X and M_Y. We can see that this metric has the desired similarity property. Consider two machines A and B, such that for every program P the ratio between their respective execution times is always a constant K. From (3) we can see that, in this situation, their pershape distance is zero. Conversely, if the distance between two machines is very large, then their pershapes are very different and the spectrum of benchmark execution time ratios is large, independently of whether the machines have a similar average performance or not. A more formal discussion about the properties of this metric can be found in ref. 18.

	Most Similar Machines					Least Similar Machines		
	machine	machine	dist.			machine	machine	dist.
001	MIPS M2000	DEC 3100	0.186		170	IBM 6000/530	Sun 3/260	1.758
002	VAX 8600	VAX 785 f77	0.229		169	IBM 6000/530	Sun 3/50	1.669
003	Sun 4/200	Sun 4/110	0.243		168	MIPS M2000	Sun 3/260	1.658
004	Sun 3/260	Sun 3/50	0.290		167	DEC 3100	Sun 3/260	1.645
005	Sparc I	DEC 3100	0.326		166	Sparc I	Sun 3/260	1.605
006	MIPS M2000	Sparc I	0.373		165	MIPS M2000	Sun 3/50	1.598
007	Sparc I	Sun 4/110	0.378		164	CRAY Y-MP	Sun 3/260	1.583
008	Sparc I	Sun 4/200	0.391		163	DEC 3100	Sun 3/50	1.573
009	Amdahl 5880	VAX 785 f77	0.422		162	NEC SX-2	Sun 3/50	1.535
010	VAX 785 fort	VAX 785 f77	0.426		161	CRAY Y-MP	Sun 3/50	1.527
011	IBM 6000/530	DEC 3100	0.442		160	Sparc I	Sun 3/50	1.521
012	IBM 6000/530	MIPS M2000	0.445		159	VAX 785 fort	Sun 3/260	1.519
013	DEC 3100	Sun 4/110	0.453		158	NEC SX-2	Sun 3/260	1.477
014	Motorola 88k	Sun 4/200	0.456		157	VAX 785 fort	Sun 3/50	1.443
015	DEC 3100	VAX 785 fort	0.462		156	IBM 3090/200	Sun 2/260	1.438
016	MIPS M2000	Sun 4/200	0.462		155	IBM 3090/200	Sun 3/50	1.425
017	MIPS M2000	Sun 4/110	0.468		154	Sun 4/110	Sun 3/260	1.399
018	IBM 6000/530	Motorola 88k	0.485		153	Motorola 88k	Sun 3/260	1.370
019	DEC 3100	Sun 4/200	0.495		152	Sun 4/200	Sun 3/260	1.344
020	VAX 8600	VAX-11/780	0.499		151	Sun 4/110	Sun 3/50	1.326

Table 11.3: Pairs of machines with the smallest and largest pershape distance.

Pershapes and pershape distance can be used to compare and understand machine performance. For example, 1) normalized pershapes allow us to identify, relative to a particular machine, the strengths and limitations of different machines. 2) We can observe the performance evolution of different implementations of the same architecture, and measure changes over time. 3) We can measure how advances in technology affect the way machine designers allocate resources to improve performance and identify those dimensions that are given more importance. 4) Pershapes metrics can be defined and used to cluster machines. In particular, machine similarity is a measurement of the potential variability of benchmark results between two machines.

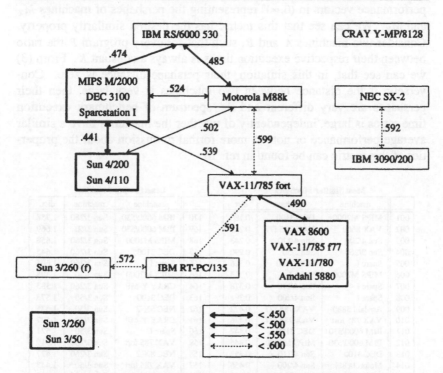

Figure 11.3: A clustering diagram based on machine distances. In order to make the figure more understandable, machines with very small pershape distances are depicted as a single unit.

In Table 11.3 we give the list of the twenty smallest and twenty largest pershapes distances for the pairs of machines in Fig. 11.2. As expected, machines with similar characteristics, like the VAXes, show small distances, but there are also other interesting results. The Sparcstation I has a smaller distance to the DEC 3100 (a MIPS Co. RC2000 and RC2010 based machine) and MIPS M/2000 than to the any of the other two Sun 4 models.

Figure 11.3 shows the clustering of machines with respect to their distances. The distance between clusters represent the average of all distances between pairs of elements in the clusters. The figure illustrates the close similarity of modern RISC microprocessors. The supercomputers have performance distributions that are significantly different than those of the other machines; only the NEC SX-2 and the IBM 3090/200 have a relatively small distance. A more complete exposition of pershapes and their metric can be found in ref. 18.

11.7. PROGRAM CHARACTERIZATION

One of our main arguments is that program characterization is important for CPU performance evaluation. Knowing the static and dynamic statistics of programs is what explains why a machine executes some programs faster, but others more slowly. There is an underlying assumption, by people who use benchmarks, that large real programs with a long execution times are always better for benchmarking purposes. In some cases this assumption is false and conclusions drawn from these programs can be quite misleading. It is by making a detailed analysis of program execution that we discover what it is that a program measures and how to interpret its benchmark results.

Our program characterizer works similarly to most execution profilers[14]. It does a static analysis of the source code at the basic block level and instruments the program with counters to measure, at run time, the number of times that each block is executed. Compiling the instrumented program and running it gives the dynamic counts. By combining the static statistics and the dynamic counts, we can compute the dynamic statistics and other interesting quantities. Merging the machine characterizations and the dynamic statistics allows us to make execution time predictions.

The programs we use in this chapter to illustrate program characterization come from the SPEC89 suite[24]. The SPEC (System Performance Evaluation Cooperative) benchmarks are the results of an effort from computer manufacturers to assemble a suite of realistic and interesting benchmarks. These programs have been selected from a large sample of public domain applications. The 1989 benchmark suite consists of ten programs, 6 in Fortran and 4 in C, covering scientific and system applications. The suite was extended in 1992 and 1995 to include a larger set of programs and also suites for measuring OS and I/O performance; some programs with unsuitable behavior were dropped. Other efforts of SPEC have focused on the development of a methodology for measuring and reporting benchmark results. A major contribution of the SPEC group has been in raising the level of discussion between machine manufacturers with respect to machine performance. However, the SPEC people have focused mainly in devising better ways to measure machine performance, and have given little attention to the problem of explaining how and why machine performance is achieved, although this situation appears to be

changing[24]. In this chapter, and in refs. 19-20, we present some results of
our analysis of the SPEC Fortran benchmarks; Gee et al.[10] report on cache
miss ratio for the SPEC92 benchmarks.

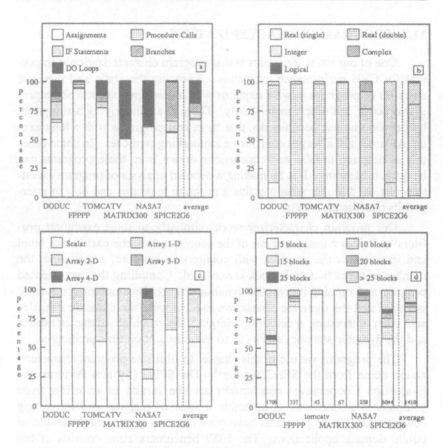

Figure 11.4: Dynamic statistics for the SPEC benchmarks. In 4d, the number at the
bottom of each bar represents the basic blocks in the benchmark.

Figure 11.4 shows machine-independent dynamic statistics for the
SPEC Fortran programs. The statistics we present are for the complete
program, but in our system it is equally easy to limit them to individual
subroutines or arbitrary groups of basic blocks. Figure 11.4 contains four
graphs, each focusing on some particular aspect of the programs: the type
of statements executed, the data type and precision of arithmetic and logical operations, the structure of the operands, and the distribution of blocks

executed. From Fig. 11.4a, we see that the most executed statements are assignments and DO LOOPs. However, the number of branches in SPICE2G6 is inordinate compared with the other benchmarks. The graph for arithmetic and logical operations (11.4b) is dominated by double precision floating point arithmetic, with the exception of SPICE2G6 (using model GREYCODE) in which integer arithmetic represents more than 80% of all operations. As we explain below, this is more a feature of the model (input data) used than the 'normal' behavior of SPICE2G6.

Very interesting is graph 11.4d, the one giving the distribution of operations with respect to basic blocks executed. At the bottom of each bar we indicate the number of basic blocks in each program. In programs FPPPP, TOMCATV, and MATRIX300, five blocks contain more than 85% of all operations. In fact for MATRIX300, a single basic block, containing a single statement, accounts for 99.8% of all operations. The situation is not much better for NASA7 or SPICE2G6. SPICE2G6 represents a good example of a large program (18000 lines in 6044 blocks) that has a very large execution time (> 20000 seconds in a VAX-11/780), but where only five basic blocks account for more than 50% of all operations; the most popular four basic blocks contain less than 10 lines. From the number of blocks, we see that TOMCATV and MATRIX300 are very small programs (43 and 67 blocks); this is corroborated by the number of lines, 183 and 149 respectively (excluding comments). These results underline the importance of taking into account, especially when we compute statistics from benchmark results, that some programs measure only very few things, while others exercise more aspects and units of each machine. This is evident in the SPEC results for the Stardent 3010[25]. For this machine the SPECratio results, excluding MATRIX300, range from 14.7 to 62.9[†]. However, the SPECratio for MATRIX300 is 108.5. It is risky to draw conclusions from these numbers without knowing what each benchmark measures.

When we analyze benchmarks in terms of a large number of characteristics, it is always useful to be able to classify them into different groups or clusters. A similar approach as the one we described for machine similarity çan be used to classify benchmarks, although the properties we want to impose on the the metric are different. In refs. 19-20 we define a metric to benchmark similarity. In this chapter we will present a more graphical approach using Chernoff faces to identify similar programs[4]. A Chernoff face is a graphical method of mapping multidimensional data to facial features[‡]. It has been observed that Chernoff

---

[†] The SPECratio is defined as the ratio between the execution time of the measured machine and the execution time of the VAX-11/780.

[‡] Standard Chernoff faces do not have hair and can be used to represent up to 20

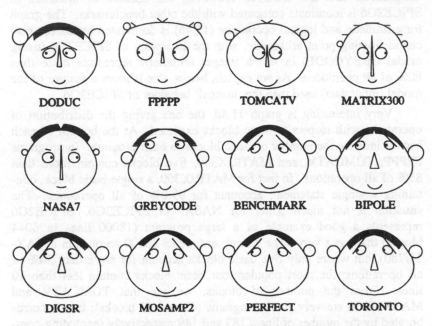

Figure 11.5: Chernoff faces for the SPEC'89 Fortran benchmarks and six additional circuits for SPICE2G6. Each face is determined by 23 program characteristics.

faces are very useful for clustering, because humans have a very good ability to recognize faces. In Figs. 11.5-6 we show examples of Chernoff faces. Each of the faces consists of 23 characteristics which include many of the measurements we showed in Fig. 11.4. In Fig. 11.5 the six faces in the upper row are for the SPEC benchmarks (SPICE2G6 is labeled as GREYCODE; the name of the circuit being analyzed). On the lower portion of the figure we show Chernoff faces corresponding to six additional circuit models for SPICE2G6. If we focus only the first row we see that there are strong similarities between TOMCATV and MATRIX300, and to a lesser degree between these two and FPPPP and NASA7. On the other hand DODUC and GREYCODE show little resemblance to any of the other four faces.

Interesting observations can be made on the Chernoff faces for the seven models of SPICE2G6. These are represented by the lower faces plus GREYCODE. From these we immediately notice that five of the

parameters; we have made them hairy to accommodate 3 more parameters.

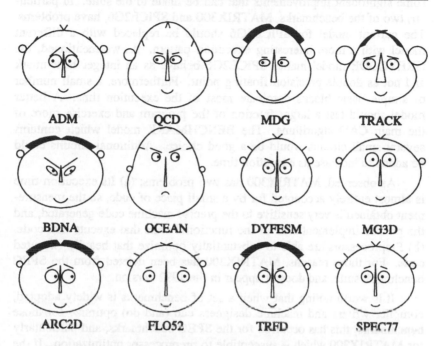

Figure 11.6: Chernoff faces for the Perfect benchmarks. Each face is determined by
23 program characteristics.

seven faces are extremely similar. Only BIPOLE and GREYCODE look
different, but even here the difference from GREYCODE to the other
models is more conspicuous. We already mentioned that SPICE2G6
using GREYCODE is dominated by integer arithmetic and not by double
precision floating point operations. Although it is not clear from the
Chernoff faces, the other circuit models are dominated by double precision arithmetic, as users of this benchmark normally assume. Another
difference between the seven circuits is that GREYCODE requires two
order of magnitude more time to execute. We have looked in some detail
at the execution of GREYCODE and found that what this model is
measuring is more the performance limitations of the data structures in
SPICE2G6 when it analyzes large circuits than the behavior of CAD algorithms. Additional Chernoff faces for the Perfect Club benchmarks are
given in Fig. 11.6. These clearly show more variability amongst the programs than those in Fig. 11.5.

Our study of the SPEC89 Fortran benchmarks indicates that there are some significant improvements that can be made to the suite. In particular, two of the benchmarks, MATRIX300 and SPICE2G6, have problems. The current model for SPICE2G6 should be replaced with a different model with a more interesting execution pattern. As we mentioned, the GREYCODE model makes SPICE2G6 behave as an integer benchmark and not as double precision floating point. Furthermore, a small number of simple basic blocks dominate most of the execution time. A better model should test a larger fraction of the program and exercise more of the main CAD algorithms. The BENCHMARK model which contains several small circuits would be a good choice. Additional circuits could be added to increase its execution time.

As observed, MATRIX300 has two problems: (a) Its execution time is almost entirely accounted for by a small piece of code, so the measurement obtained is very sensitive to the precise machine code generated, and the precise implementation of the functional units that execute that code. (b) Preprocessors are able to substantially optimize that heavily executed code. For those reasons, MATRIX300 has been deleted from the SPEC benchmark suite, and doesn't appear in the 1992 version.

It is worth noting that when a set of benchmarks is widely adopted, compiler writers and machine designers can (and do) optimize for those benchmarks; this has occurred for the SPEC benchmarks, and particularly for MATRIX300 which is susceptible to preprocessor optimization. If the benchmarks are too simple or do not represent real workloads, the apparent improvements will not translate in actual gains for users' applications. The best way to prevent this is by knowing what each benchmark measures and replacing those that are inadequate.

11.8. EXECUTION TIME PREDICTION

Any execution time model needs to be able to make accurate execution time predictions for real programs in order to be considered credible and useful. It is only by comparing the predictions to actual measurements that we can quantify the accuracy of the model, region of validity, and robustness. Once we have validated the model we can use it to predict execution times, to study how sensitive the execution time is to changes in the workload, to assess the impact of new algorithms, etc.

We have predicted and compared execution time predictions for a large set of programs and machines, and we have found that our predictions are generally quite accurate. For a sample of more than 244 machine-program combinations, using 20 machines and 28 programs, we

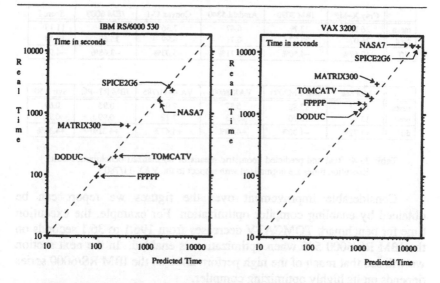

Figure 11.7: Comparison between real execution times and predictions for the IBM RS/6000 530 and VAX 3200.

have found that 55% of the predictions lie within 10% of the real execution time, 80% within 20%, and 94% within 30%. In Fig. 11.7 we show two graphs, those for two machines (IBM RS/6000 530 and VAX 3200) and the six SPEC Fortran programs, which present a comparison between actual execution times and predictions. The execution times range from less than 100 seconds to more than 15000.

Benchmark results are often reduced to a single number representing the absolute performance of the machine or its relative performance with respect to a baseline machine. A common approach in benchmarking is to compute the geometric mean of the execution times normalized by the results of the VAX-11/780. This is the approach taken in the SPEC benchmarks. Here we will show, using the same performance metric, how close the single number performance estimates computed using our predictions and those obtained from real execution times are.

In Table 11.4 we present both the actual and predicted geometric means for twelve machines ranging from the CRAY X-MP to the Sun 3/50. All programs were executed in scalar mode and without optimization, which accounts for the fact that many of our figures are smaller than reported SPECmarks. In all cases the difference between the predicted and real value is less than 6%; the average is +0.22%. Clearly, our methodology yields good predictions of relative performance.

	Cray X–MP	IBM 3090	Amdahl 5840	Convex C-1	IBM 6000	Sparc I
mean	26.25	33.79	6.47	7.36	16.29	11.13
pred	26.07	32.27	6.71	6.99	15.69	10.58
diff	+0.69%	–4.50%	+3.71%	–5.03%	–3.68%	–4.94

	Mot 88k	MIPS M2000	VAX 8600	VAX–11/785	IBM RT–PC	Sun 3/50
mean	14.24	13.88	5.87	2.01	0.95	0.69
pred	15.34	13.70	5.63	2.12	0.99	0.72
diff	+7.72%	–1.30%	–4.09%	+5.47%	+4.21%	+4.35%

Table 11.4: Real and predicted geometric means of normalized benchmark results.
Execution times are normalized with respect to the VAX-11/780.

Considerable improvement over the figures we report can be obtained by enabling compiler optimization. For example, the execution time for benchmark TOMCATV decreases from 196.1 to 36.1 seconds on the IBM RS/6000 530 when optimization is enabled. In the next section we will see that much of the high performance on the IBM RS/6000 series depends on its highly optimizing compiler.

11.9. Compiler Optimization

Thus far, we have presented our linear model as if a given source program were simply and directly translated into the corresponding machine code. In reality, compilers optimize the code, even when the optimizer is nominally inactive (optimization level 0). Including the effect of optimizations in our analysis is a difficult problem; we summarize some of the issues and our results in this section.

The problem of evaluating the effects of compiler optimizers can be broken down into three subproblems. 1) The experimental detection of which optimizations can be applied by the optimizer –i.e. what can the compiler optimizer do. 2) The measurement of the performance improvement that a particular optimization will produce in a program. 3) The measurements of the possible optimizations present in the source code. Each is discussed below.

The first point is similar to machine characterization, but instead of measuring the performance of some operation, we are interested in detecting which optimizations can be applied by the compiler and in which cases. This may appear as an easy task, but unfortunately, in many compilers, optimizations are implemented ad-hoc; e.g. a given transformation may be used for only one data type and not another, although the transformation is type independent[12].

compiler	constant folding	common subexpr elim	code motion	copy propagation	dead code elimination
Ultrix F77 1.1	no	partial	marginal	partial	no
Mips F77 1.21 -O2	partial	yes	yes	partial	yes
Mips F77 1.21 -O1	marginal	yes	no	marginal	no
Sun F77 -O3	marginal	yes	yes	no	yes
Sun F77 -O2	marginal	yes	yes	no	partial
Sun F77 -O1	no	no	no	no	no
Ultrix Fort 4.5	yes	yes	yes	yes	yes
Amdahl F77 2.0	no	no	no	no	no
CRAY CFT77 4.0.1	yes	yes	yes	yes	yes
IBM XL Fortran 1.1	yes	partial	yes	partial	yes
Motorola F77 2.0b3	marginal	yes	yes	no	no

Table 11.5: Summary of the effectiveness of compilers in applying local optimizations (1 of 2).

compiler	strength reduction	address calculation	inline substitution	loop unrolling
Ultrix F77 v.1	partial	marginal	no	no
Mips F77 1.21 -O2	yes	yes	marginal	yes
Mips F77 1.21 -O1	no	yes	no	no
Sun F77 -O3	partial	marginal	no	yes
Sun F77 -O2	partial	no	no	yes
Sun F77 -O1	no	no	no	yes
Ultrix Fort 4.5	yes	yes	no	no
Amdahl F77 2.0	no	no	no	no
CRAY CFT77 4.0.1	yes	yes	yes	yes
IBM XL Fortran 1.1	yes	yes	partial	yes
Motorola F77 2.0b3	partial	no	no	no

Table 11.6: Summary of the effectiveness of compilers in applying local optimizations (2 of 2).

We have written a benchmark program to detect scalar optimizations in Fortran compilers; see ref. 1 for a discussion of compiler optimization. It contains experiments that check whether individual optimizations can be detected inside a basic block (local) and across basic blocks (global). We also test separately that the optimizations work on integers, reals, and mixed (integer and real) expressions. All optimizations we detect are machine independent.

In Tables 11.5 and 11.6 we present a summary of the results found for eight Fortran compilers for local optimizations; see refs. 19, 21 for further discussion. In addition to 'no' or 'yes', we also use the terms 'marginal' and 'partial'. Marginal is used when the optimization is only detected in some special cases. Partial means that the optimization is detected in most, but not all situations. For example, if an optimization is applied on expressions containing integers or reals, but not when both are present, we use 'partial'. If it is only applied on one data type we use 'marginal'.

Inline substitution can be used to illustrate how tricky is to evaluate an optimizing compiler. Calls to leaf procedures, those that do not call other procedures, can be eliminated, and potential optimization exposed, by inserting the callee's code at the point of call, after making a proper substitution for the formal parameters. In Table 11.6 we see that our test detected that three compilers have some ability to inline procedures, but only the CFT77 compiler takes full advantage of it. In the case of MIPS f77 1.21, the compiler does not perform an actual inline substitution. The only transformation done is that the compiler does not use a new stack frame for the leaf procedure, but instead execution is carried out on the caller's frame. In contrast, a real inline substitution is done by IBM XLF 1.1, but here the insertion of unnecessary extra code obscures optimizations that inlining should have exposed. Only the CFT77 compiler was abled to detect all optimizations present after proper inlining.

The second subproblem mentioned above deals with quantification of the performance benefits of individual transformations. Several studies have tried to measure the performance effect of some algorithms for code improvement[7,5]. Most have been carried out in the process of engineering an optimizing compiler and performance evaluation has not been the main driving force; ref. 16 is an exception. In addition, some of the studies have used a set of small programs with a very regular structure like matrix multiplication, FFT, Baskett puzzle, etc, in which the execution time is significantly reduced by applying one or two transformations.

The third subproblem refers to the amount of optimizable code present in real applications. Detecting which optimizations are done by optimizers and measuring their performance effects is only part of the problem; we also need to know the extent to which programs contain optimizable code. The measurement of optimization opportunities in programs could potentially be done by modifying an existing highly optimizing compiler, but of course we could then only measure the optimizations that compiler was able to find.

Although the last two problems are important, and each will require a significant effort to solve them, in our research we have adopted a different strategy. As one part of our work on this problem, we have used a suite of large scientific programs to measure the average improvement produced by different optimizing compilers, and have also used these

compiler	average	std. dev.
Ultrix F77 1.1	.7781	.1295
Mips F77 1.21 -O2	.5117	.1898
Mips F77 1.21 -O1	.8420	.0564
Sun F77 -O3	.5230	.2421
Sun F77 -O2	.5537	.2038
CRAY CFT77 4.0.1	.5640	.2933
IBM XL Fortran 1.1	.2849	.1381
Motorola F77 2.0b3	.6869	.1404

Table 11.7: Average decrease (the ratio between optimized and non-optimized execution time) produced by several optimizers and levels of optimization observed on the Perfect Club benchmarks.

results to investigate the amount of correlation found between optimizers. Table 11.7 gives results with respect to the average change (ratio of final to initial) in execution time obtained when optimization is enabled. We also give the standard deviation. It is important to realize that a larger reduction in execution time does not necessarily mean that the optimization is better, but only that the improvement to the original object code is larger; the original code may have been very poor.

An interesting question to consider is whether different compilers correlate in their ability to improve the execution time of individual programs. First, recall that we can predict execution times for nonoptimized programs. Now, if we find that there exists a significant correlation between two optimizers, then knowing how much one optimizer reduces execution time will allow us estimate the reduction by the other optimizer. Thus, this give us a way of predicting execution times before and after optimization. In Fig. 11.8 we show, for those compilers in Table 11.7, graphs of execution time reductions for all pairwise combinations of compilers. We include the best linear fit to the set of points.

As expected there is a positive correlation between all optimizers; on the average more improvement by one compiler means more improvement in the other. All the coefficients of correlation, except three, are greater than .6300. However only those for graphs 1, 2, 3, 5, 6, 11, and 12, are statistically significant at a level of 0.025. An analysis of the code produced by the optimizers on our suite shows that the optimizations having the most effect are: 1) Reducing the number of loads and unnecessary stores by keeping temporaries in registers (see Fig. 11.9). 2) The elimination of expensive address computations by computing a base address before the first iteration and updating it with an add in all subsequent iterations. 3) Moving invariant code outside the innermost loops.

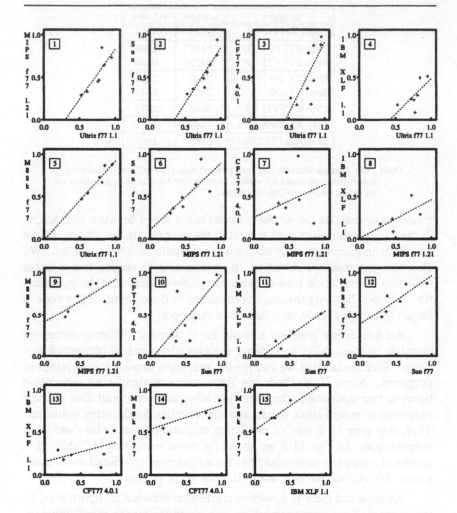

Figure 11.8: Correlation between execution time improvements of various optimizing compilers. Each graph includes the best linear fit.

Several problems arise when we deal with compiler optimizers and
attempt to measure their effects. The first is that most optimizations cannot be measured in isolation. It is common that the opportunity to apply
some optimization is the result of a previous transformation, and in some
cases the first transformation may not necessarily improve the execution
time of the program. A second problem is machine-dependent optimizations. Proposing a general framework for the evaluation of optimizers

hampers our ability to evaluate machine-dependent optimizations. Most of these optimizations do not have an equivalent in other architectures, and it may be argued that these transformations are not really optimizations of the source code, but deal with the efficient use of the machine resources. An example is the IBM RS6000, which has a 'multiply-add' instruction. More generally, the same problem occurs in any superscalar or VLIW machine in which there is parallelism at the functional unit level.

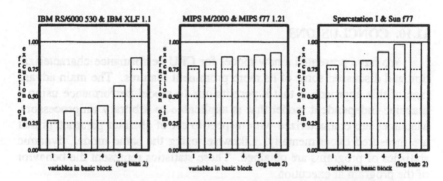

Figure 11.9 The effect of register allocation for three different machine-compiler combinations. The numbers on the x-axis are the logarithm base 2 of the number of variables that the optimizer must consider as candidates for registers.

A machine-dependent optimization that we have attempted to measure is register allocation. The speedup obtained by this optimization depends not only on the quality of the algorithm, but also on the number of registers and the time differential between using a register as opposed to loading the value from the cache or memory. Even when the program does not show opportunities for optimization, the optimizer improves the code by doing peephole optimization, so it is not possible to measure the register allocation speedup in isolation. Nevertheless, it is possible to detect the range of improvement derived from this optimization. We have written tests where the only opportunities for improvement are register allocation and peephole optimization. By increasing the variables referenced inside a loop, while keeping everything else constant, we make it more difficult for the optimizer to keep all variables in registers. As a smaller fraction of variables are kept in registers, the optimized execution time tends to get closer to the non-optimized version.

We present results for three machine-compiler pairs in Fig. 11.9. Each graph shows how much optimization changes the execution time as a function of the logarithm of the number of distinct variable referenced inside the main loop. The results indicate that the gap between optimized and non-optimized execution times decreases as the number of variables increases; this is clearest for the IBM XLF compiler. The IBM RS/6000 contains in addition to 32 integer and 32 floating point registers, register renaming to prevent stalling, and a very effective register allocation algorithm. The MIPS f77 and Sun f77 compilers show similar improvement from register allocation, but the effect of increasing the 'size' of the basic blocks is more steep for the Sun f77 optimizer.

11.10. CONCLUSIONS

We have presented a new model for CPU performance characterization and discussed some of its more prominent features. The main advantage of this approach is that it permits one to model performance using a machine-independent model that is applicable to arbitrary uniprocessors. Machines are characterized with respect to set of abstract parameters that are measured experimentally. Likewise, using the same model, dynamic statistics of programs are obtained. These statistics represent the behavior of the program at execution.

By combining the machine and program characterizations, it is possible to predict with good accuracy execution times for arbitrary large programs over a wide spectrum of machines. In addition, it is possible to combine individual measurements into a set of reduced parameters that better represent hardware or software components. This permits us to identify the performance strengths and weaknesses of machines. We have defined abstract concepts like machine and program similarity that further increase our insight about how machines and programs behave. In ref. 21 this work is extended to optimizing compilers and in ref. 22 we examine the effect of the memory hierarchy on run times. Work in ref. 27 extends some of our results to parallel and vector machines. Overall, we believe that we have developed a powerful and useful methodology for measuring and evaluating machine performance, and for characterizing benchmark programs.

11.11. ACKNOWLEDGEMENTS

The material presented here is based on research supported principally by NASA under grant NCC2-550, and also in part by the National Science Foundation under grants MIP-8713274, MIP-9116578, CCR-9117028, and CCR-9308981, by the State of California under the MICRO program, and by Intel Corporation, Apple Computer Corporation, Sun

Microsystems, Digital Equipment Corporation, Philips Laboratories/Signetics, International Business Machines Corporation, and Mitsubishi Electric Research Laboratories.

11.12. BIBLIOGRAPHY

1. A.V. Aho, R. Sethi, and J.D. Ullman, *Compilers, Principles, Techniques, and Tools*, Addison-Wesley, Reading, Mass., 1986.

2. D.F. Bacon, S.L. Graham, and O.L. Sharp, "Compiler Transformations for High-Performance Computing", *Computing Surveys*, November 1994, pp. 343-420.

3. D.H. Bailey and J.T. Barton, "The NAS Kernel Benchmark Program", NASA Technical Memorandum 86711, August 1985.

4. H. Chernoff, "The Use of Faces to Represent Points in k-Dimensional Space Graphically", *J. of the American Statistical Association*, Vol. 68, No. 342, June 1973, pp. 361-368.

5. F. Chow, *A Portable Machine-Independent Global Optimizer - Design and Measurements*, Ph.D. dissertation, Technical Report 83-254, Computer Systems Laboratory, Stanford University, December 1983.

6. R.M. Clapp, L. Duchesneau, R.A. Volz, and T. Schultze, " Toward Real-Time Performance Benchmarks for ADA", *Communications of the ACM*, Vol. 29, No. 8, August 1986, pp. 760-778.

7. J. Cocke and P. Markstein, "Measurement of Program Improvement Algorithms", *Research Report*, IBM, Yorktown Heights, RC 8110, July 1980.

8. G. Cybenko, L. Kipp, L. Pointer, and D. Kuck, "Supercomputer Performance Evaluation and the Perfect Benchmarks", University of Illinois Center for Supercomputing R&D Technical Report 965, March 1990.

9. J.J. Dongarra, J. Martin, and J. Worlton, "Computer Benchmarking: paths and pitfalls", *Computer*, Vol. 24, No. 7, July 1987, pp. 38-43.

10. J. Gee, M. Hill, D. Penvmatikatos, and A.J. Smith, "Cache Performance of the SPEC Benchmark Suite" *IEEE MICRO*, Vol.13, No.4, August, 1993, pp. 17-27.

11. D. Hinnant, "Accurate Unix Benchmarking: Art, Science or Black Magic?", *IEEE MICRO*, October, 1988, pp. 64-75.

12. D.S. Lindsay, "Methodology for Determining the Effect of Optimizing Compilers", *CMG 1986 Conference Proceedings*", Las Vegas, Nevada, pp. 379-385, December 9-12, 1986.

13. D.N. Pnevmatikatos and M.D. Hill, "Cache Performance of the Integer SPEC Benchmarks on a RISC, *Computer Architecture News*, Vol. 18, No. 2, June 1990, pp. 53-68.

14. L.R. Powers, "Design and Use of a Program Execution Analyzer", *IBM Systems Journal*, Vol. 22, No.3, 1983, pp. 271-292.

15. W. Price, "A Benchmark Tutorial", *IEEE MICRO*, October, 1989, pp. 28-43.

16. S. Richardson and M. Ganapathi, "Interprocedural Optimization: Experimental Results", *Software–Practice and Experience*, Vol. 19, No. 2, pp. 149-170, February 1989.

17. R.H. Saavedra-Barrera, "Machine Characterization and Benchmark Performance Prediction", University of California, Berkeley, Technical Report No. UCB/CSD 88/437, June 1988.

18. R.H. Saavedra-Barrera, A.J. Smith, and E. Miya, "Machine Characterization Based on an Abstract High Level Language Machine", IEEE Transactions on Computers, special issue on Performance Evaluation, December, 1989, 38, 12, pp. 1659-1679.

19. R.H. Saavedra-Barrera, *CPU Performance Evaluation and Execution Time Prediction Using Narrow Spectrum Benchmarking*, Ph.D. Thesis, UC Berkeley, Tech. Rept. No. UCB/CSD 92/684, February 1992.

20. R.H. Saavedra and A.J. Smith, "Analysis of Benchmark Characteristics and Benchmark Performance Prediction", Technical Report UCB/CSD-92-715, December, 1992.

21. R.H. Saavedra and A.J. Smith, "Benchmarking Optimizing Compilers", *IEEE Trans. on Software Eng.*, Vol. 21, No. 07, July 1995, pp. 615-628.

22. R.H. Saavedra and A.J. Smith, "Measuring Cache and TLB Performance", *IEEE Trans. on Computers*, Vol, 44, No. 10, October 1995, pp. 1223-1235.

23. J.P. Singh, W.-D. Weber, and A. Gupta, "SPLASH: Stanford Parallel Applications for Shared Memory", *Comp. Arch. News*, Vol.20, No.1, March 1992, pp. 5-44.

24. SPEC, *"SPEC Newsletter: Benchmark Results"*, Vol. 1, Issue 1, Fall 1989.

25. SPEC, *"SPEC Newsletter: Benchmark Results"*, Vol. 2, Issue 2, Spring 1990.

26. SPEC, *Technical Manual*, Rev.1.1, 1992.

27. S. Von Worley and A.J. Smith, "Microbenchmarking and Performance Prediction for Parallel Computers", Technical Report UCB/CSD-95-873, May, 1995.

28. R. Weicker, "Dhrystone: A Synthetic Systems Programming Benchmark", CACM, 27, 10, October, 1984, pp. 1013-1030.

29. S.C. Woo, M. Ohara, E. Torrie, J.P. Singh, and A. Gupta, "The SPLASH-2 Programs: Characterization and Methodological Considerations", *Proc. of the 22nd Anual Int. Symp. on Comp. Arch.*, Santa Margherita Ligure Italy, June 22-24 1995, pp. 24-36.

30. J. Worlton, "Understanding Supercomputer Benchmarks", *Datamation*, September 1, 1984, pp. 121-130.

CHAPTER 12

EVALUATION AND DESIGN
OF BENCHMARK SUITES

Jozo J. Dujmović

12.1. INTRODUCTION

Benchmark suites are most frequently designed for industrial evaluation of competitive computer systems and networks. Examples of such benchmark suites include SPEC[1], TPC[2], GPC[3], PERFECT Club Benchmarks[4], AIM benchmarks[5], and others. In addition to benchmark suites sponsored by consortia of computer industry there are various collections of benchmarks designed by research organizations, companies, computer magazines, and individuals for benchmarking specific hardware and software systems. Examples of such benchmark suites include database benchmarks[6], supercomputer benchmarks (Livermore loops[7], NAS Parallel Benchmarks[8], Lisp benchmarks[9], Prolog benchmarks[10], and many others[11]. In the majority of cases benchmark workloads are selected from a specific set of frequently used real workloads. The selection process is usually aimed at simultaneous satisfaction of two goals: (1) benchmark workloads should be a good functional representative of a given universe of real workloads, and (2) benchmark workloads should yield the same distribution of the utilization of system resources as real workloads. Due to the absence of specific quantitative methods for benchmark suite design, these goals primarily serve as guidelines in an intuitive workload selection process. Such a process cannot include proofs of the extent to which the goals are satisfied,

287

and frequently yields relatively low reliability of benchmark results and excessive cost of benchmarking.

This chapter will propose quantitative methods for evaluation and design of benchmark suites which can increase the reliability and decrease the cost of benchmarking of computer systems and networks. These methods are based on appropriate white-box and black-box workload difference models. The models are then used to define benchmark suite performance criteria, and to develop techniques for benchmark suite evaluation and design. The proposed design techniques are based on a quantitative optimization of benchmark suite performance criteria.

12.2. WORKLOAD DIFFERENCE CONCEPTS

Let us consider a set of n workloads, denoted $B_1, \ldots, B_n, n > 1$. We assume that these workloads are used as benchmark programs for performance measurement and comparison of m competitive computer systems, $S_1, \ldots, S_m, m > 1$. In this context we are interested in developing quantitative models of difference and similarity between each pair of workloads. The difference between B_j and B_k will be denoted $d(B_j, B_k)$, and the similarity will be denoted $s(B_j, B_k)$. The first step in the development of workload difference models consists of introducing the concepts of minimum and maximum difference between workloads.

A *program space* is a space where points represent programs (or some other more complex computer workloads). Differences between computer workloads can be expressed as distances between points in the program space. The proximity of two points in the program space denotes the similarity of corresponding computer workloads. Therefore, similar workloads are represented as clusters of points in the program space.

The difference between workloads $d_{jk} = d(B_j, B_k)$ always satisfies the mathematical conditions of non-negativity ($(\forall j, k)\ d_{jk} \geq 0$), non-degeneracy ($d_{jk} = 0$, for $j = k$, and $d_{jk} > 0$, for $j \neq k$), and symmetry ($d_{jk} = d_{kj}$). These conditions are known as *semimetric*.

In benchmarking it is important to note that a given computer workload can be expressed using several equivalent source program forms, and all such forms must be interpreted as the same workload. In other words, $d_{jk} = 0$ simply implies that B_j and B_k are two equivalent forms of the same workload. For example, in C, the *for* loop

```
for(expr_1; expr_2; expr_3) statement;
```

is equivalent to the *while* loop

```
expr_1; while (expr_2) { statement; expr_3; }
```

After compiling and linking, both loops should generate the same executable code. So these are equivalent source forms of the same workload. Similarly, many benchmark workloads have a loop form

```
for(i=0; i<K; i++) benchmark_workload();
```

where K denotes a scaling factor used to adjust the total run time according to the desired accuracy and convenience of measurement. We assume that the nature of workload does not depend on the value of K, and that the above *for* loop, for any $K > 0$, is equivalent to the loop body, benchmark_workload().

Let us now address the concept of maximum difference between benchmark workloads. If the difference is not limited, then some workloads might have an "infinite difference." Unfortunately, it is not easy to find a reasonable intuitive interpretation of such a difference. In addition, it is not possible to use fractions of the maximum difference, and it is not possible to define similarity as a simple linear complement of difference. On the other hand, if we assume that the difference between benchmark workloads is limited, and that the maximum difference is d_{max}, then any specific model of difference between workloads will include the concept of "completely different workloads," where $d(B_j, B_k) = d_{max}$. For each specific difference model the completely different workloads must have an easily acceptable intuitive interpretation which supports the credibility of the model. All differences can now be conveniently interpreted as fractions of d_{max}, and the similarity between workloads can be defined as a complement of difference:

$$s(B_j, B_k) = d_{max} - d(B_j, B_k) .$$

Therefore, it is reasonable to make the following fundamental assumption:

$$0 \leq d(B_k, B_k) \leq d_{max} < +\infty ,$$

which we call *the concept of limited difference*. For most applications it is convenient to adopt $d_{max} = 1$, or $d_{max} = 100\%$.

Let us now show that all workload difference models fundamentally depend on the purpose of benchmarking. In other words, we cannot expect to have a single workload difference model. Each specific benchmarking problem can yield a corresponding workload difference model. This is consistent with the traditional classification of workload models which differentiates physical models (which are system dependent and based on utilization of hardware and software), virtual models (reflecting a programmer's viewpoint of the use of logical resources), and functional models (which are system independent and application oriented).

Benchmarking is most frequently performed to generate data which can be used for comparison and ranking of competitive systems. Comparison is usually based either on measured run times $(t_1, \ldots, t_m)$, or throughputs $(x_1, \ldots, x_m)$. If ε_t and ε_x denote maximum expected measurement errors for times and throughputs respectively, then the notation used for ranking competitive systems can be defined as follows:

$S_j \succ S_k$: S_j is better than S_k, i.e. $t_j < t_k - \varepsilon_t$, or $x_j > x_k + \varepsilon_x$

$S_j \succeq S_k$: S_j is better than or equivalent to S_k, i.e. $t_j \leq t_k - \varepsilon_t$, or $x_j \geq x_k + \varepsilon_x$

$S_j \sim S_k$: S_j is equivalent to S_k, i.e. $|t_j - t_k| < \varepsilon_t$, or $|x_j - x_k| < \varepsilon_x$.

If the only purpose of benchmarking is to provide a ranking $(S_1 \succeq S_2 \succeq \ldots \succeq S_m)$ of m competitive systems, then all benchmarks that yield the same ranking are equivalent. This standpoint can be used when developing difference models for workloads belonging to industrial benchmark suites such as SPEC, TPC, GPC, etc.

If the purpose of benchmarking is to measure the average utilizations of N resources of a computer system, $U_1, \ldots, U_N$, then all benchmarks yielding the same utilization distribution are equivalent, and corresponding workload difference models must be based on this equivalence. This approach is suitable if the workloads are characterized by frequency distributions of machine instructions and/or addressing modes, or various Kiviat graphs.

In the traditional "lowest bid" computer acquisition process the purpose of benchmarking is to show that the processing time (t) for performing a selected set of data processing functions is less than a given threshold value t_{max}. Consequently, all implementations of selected functions satisfying $t < t_{max}$ are equivalent, and this equivalence should affect the workload difference model. This kind of benchmarking is usually based on common system functions (edit, compile, link/load, debug, profile, etc.), or frequent business applications (sort, bank transactions, word processing, payroll, inventory control, etc.).

These examples support the conclusion that workload difference models $d(B_j, B_k)$ must be derived from the analysis of the purpose of benchmarking. If the purpose of benchmarking is to get an insight into the internal organization and operation of a computer system, then workload characterization will be based on detailed measurements of the use of all individual computer resources for a set of workloads. This approach will be referred to as the *white-box* approach to workload characterization. On the other hand, if the purpose of benchmarking is to get global performance indicators (such as throughput or response time) without any interest in the internals of computer organization and operation, then the corresponding approach to workload characterization will be referred to as the *black-box* approach.

12.3.   A WHITE-BOX WORKLOAD DIFFERENCE MODEL

Let N be the number of computer resources used by benchmark programs B_j and B_k, and let U_{ji} and U_{ki} denote utilizations, i.e. the fractions of total time the ith resource is used by programs B_j and B_k respectively. Thus, $0 \le U_{ji} \le 1$, $0 \le U_{ki} \le 1$, for $i = 1, \ldots, N$. The resources can be used either sequentially or simultaneously. For example, if a processor executes machine instructions sequentially, and U_{ji} denotes the relative frequency of the ith opcode, then obviously $\sum_{i=1}^{N} U_{ji} = 1$. On the other hand, if U_{ji} denotes the utilization of the ith resource (processor, I/O channel, disk, etc.) of a multiprogrammed machine, then we can have simultaneous activity of resources yielding $\sum_{i=1}^{N} U_{ji} \le N$. Following is a model of difference between B_j and B_k that is suitable in all cases:

$$d(B_j, B_k) = \left(\frac{\sum_{i=1}^{N} |U_{ji} - U_{ki}|}{\sum_{i=1}^{N} (U_{ji} + U_{ki})} \right)^{\alpha}, \quad \alpha > 0 .$$

The parameter α is used to adjust nonlinear features of the difference model. The values $\alpha < 1$ can be used to emphasize the importance of small differences, and the values $\alpha > 1$ to emphasize the values of larger differences. The nominal case is $\alpha = 1$. Since

$$\sum_{i=1}^{N} |U_{ji} - U_{ki}| \leq \sum_{i=1}^{N} U_{ji} + \sum_{i=1}^{N} U_{ki} ,$$

it follows that $0 \leq d(B_j, B_k) \leq 1$, as expected according to the concept of limited difference.

The limit value $d_{max} = 1$ can occur if and only if B_j and B_k use two mutually disjoint subsets of resources. In the case of monoresource benchmarks[13] (workloads that predominantly use a single resource, e.g. processor, or disk) suppose that benchmark B_j predominantly uses the jth resource. Then, the maximum difference is obtained in an ideal case where

$$\begin{aligned} U_{ji} &= 1, \quad j = i , \\ &= 0, \quad j \neq i , \quad (i = 1, \ldots, N, \quad j = 1, \ldots, N) \end{aligned}$$

yielding $d(B_j, B_k) = 1$. This ideal situation is rather unlikely in practice, particularly in cases where $U_{j1}, \ldots, U_{jN}$ denote the relative frequencies of machine instructions. Pure processor workloads are usually characterized as computational (integer-intensive, or floating point intensive), combinatorial (using predominantly tests, branches, and jumps), I/O and data transfer intensive, or a combination of these basic types. However, it is almost impossible to find workloads which use ideally mutually disjoint subsets of machine instructions so that the difference $d(B_j, B_k) = (0.5 \sum_{i=1}^{N} |U_{ji} - U_{ki}|)^{\alpha}$ is close to 1. The white-box workload difference model based on frequencies of machine instructions usually yields small values of $d(B_j, B_k)$, while larger differences occur very infrequently. This can be modified using $\alpha < 1$.

12.4. BLACK-BOX WORKLOAD DIFFERENCE MODELS

The most frequent use of benchmark suites is the measurement of execution times and throughputs for n benchmark programs executed on m different computer systems. Let $t[i, j]$ be the execution time of program B_j using the system S_i. The results of measurements are organized as a table of selected performance indicators $z[i, j]$, shown in Table 12.1. The performance indicators in this table are usually defined in one of the following ways:

Table 12.1. A typical organization of the performance indicator table

Performance Indicators		Benchmark Suite Members						
		B_1	$\ldots$	B_j	$\ldots$	B_k	$\ldots$	B_n
Competitive Systems	S_1	$z[1,1]$	$\ldots$	$z[1,j]$	$\ldots$	$z[1,k]$	$\ldots$	$z[1,n]$
	$\vdots$	$\vdots$		$\vdots$		$\vdots$		$\vdots$
	S_i	$z[i,1]$	$\ldots$	$z[i,j]$	$\ldots$	$z[i,k]$	$\ldots$	$z[i,n]$
	$\vdots$	$\vdots$		$\vdots$		$\vdots$		$\vdots$
	S_m	$z[m,1]$	$\ldots$	$z[m,j]$	$\ldots$	$z[m,k]$	$\ldots$	$z[m,n]$

execution (or response) time: $z[i,j] = t[i,j]$,

throughput: $z[i,j] = x[i,j] = 1/t[i,j]$,

relative throughput: $z[i,j] = r[i,j] = t[0,j]/t[i,j]$.

The time $t[0,j]$ corresponds to a reference system S_0 (e.g. in the case of SPEC92 and SPEC95 benchmarks, the reference systems are VAX 11/780 and SUN SPARCstation 10/40, respectively).

The measured performance indicators can have very different ranges of values. Hence, the first step in each analysis is to perform the normalization of original data. Let us first consider the case where $z[i,j] = t[i,j]$. We assume that the nature of workload does not change for loop-structured equivalent workloads. Consequently, for each benchmark program B_j the measured elapsed times in the jth column of the $t[i,j]$ table can be multiplied by any positive constant. The most useful case of such a transformation is the following:

$$T[i,j] = \frac{t[i,j]}{\max_{1 \leq i \leq m} t[i,j]} , \quad 0 < T[i,j] \leq 1 , \quad 1 \leq i \leq m .$$

This form of transformation is also valid for throughput tables:

$$X[i,j] = \frac{x[i,j]}{\max_{1 \leq i \leq m} x[i,j]} = \frac{\min_{1 \leq i \leq m} t[i,j]}{t[i,j]} , \quad 0 < X[i,j] \leq 1 .$$

In the case of the relative throughput tables the division of columns by the maximum value is also possible, but the result of such a normalization is the normalized throughput table (i.e. the normalization eliminates the effect of the reference system S_0):

$$R[i,j] = \frac{r[i,j]}{\max_{1 \leq i \leq m} r[i,j]} = \frac{\min_{1 \leq i \leq m} t[i,j]}{t[i,j]} = X[i,j] .$$

Our next step is to organize the model of difference between bench-mark workloads. This difference can be expressed as a distance between column vectors in Table 12.1. The main concept of black-box models is that the proximity of workloads causes the stochastic dependence of column vectors[14]. Accordingly, the stochastic independence corre-sponds to the maximum distance. Our first model is called *the uniform random number difference*. It is based on the idea that the maximum difference between benchmark programs B_j and B_k is obtained when their normalized elapsed times $T[1..m, j]$ and $T[1..m, k]$ behave as two sequences of independent standard uniform random numbers. This ap-proach yields the following workload difference model:

$$d_{jk} = \frac{3}{m} \sum_{i=1}^{m} |T[i,j] - T[i,k]| \,, \quad 0 < T[i,j], T[i,k] \le 1, \quad 1 \le i \le m \,.$$

Indeed, if $T[1..m, j]$ and $T[1..m, k]$ are sequences of standard uniform random numbers, then

$$\lim_{m \to +\infty} \frac{1}{m} \sum_{i=1}^{m} |T[i,j] - T[i,k]| = \int_0^1 dy \int_0^1 |y - x| dx = \frac{1}{3}$$

and therefore $\lim_{m \to +\infty} d_{jk} = 1 = d_{max}$. This value of d_{max} should be in-terpreted as a "soft limit" in the sense that d_{max} denotes the maximum difference that can occur in "regular cases" (i.e. under normal circum-stances). In accidental cases, however, the difference between programs can be greater than d_{max}. These cases would occur if the columns in Table 12.1 were negatively correlated. In all normal cases the columns are positively correlated because faster machines should simultaneously reduce the run times of all benchmark programs. Thus, independent uncorrelated columns represent the limit case, and yield the maximum regular difference d_{max}. The differences greater than d_{max} are unlikely but not impossible: from $0 < T[i,j] \le 1$ and $0 \le |T[i,j] - T[i,k]| < 1$, it follows that $0 < \frac{3}{m} \sum_{i=1}^{m} |T[i,j] - T[i,k]| < 3 = d_{MAX}$. The differences greater than d_{MAX} are not possible. So, d_{MAX} denotes a "hard limit" of workload difference: $d(B_j, B_k) < d_{MAX}$ in all cases, both regular and accidental.

This and other black-box models yield an expanded (stochastic) version of the concept of limited difference, whose general form is

$$0 \le d(B_j, B_k) \le d_{max} \le d_{MAX} < +\infty, \text{ in regular cases}$$
$$|d(B_j, B_k) - d_{max}| \le \varepsilon < 0.05 d_{max} \quad , \text{ in regular limit cases of}$$

$$d_{max} + \varepsilon < d(B_j, B_k) \leq d_{MAX} < +\infty \, , \quad \begin{array}{l} \text{maximum difference} \\ \text{in accidental cases} \\ \text{of overflow} \end{array}$$

Since the maximum regular difference d_{max} is related to the stochastic independence of the columns of performance indicators, and the size of columns is limited by the number of available systems, it follows that d_{max} is computed with inevitable small random fluctuations whose range is denoted ε (normally $\varepsilon < 0.05 d_{max}$). For some difference models the concept of limited difference yields a unique limit value, i.e. $d_{max} = d_{MAX}$, as for the white-box model. For other models we have two limit values (soft and hard): $d_{max} < d_{MAX}$, as for black-box models. The soft limit d_{max} can be considered a regular limit value of the maximum difference between workloads, and we assume $d_{max} = 1$. In unlikely cases where $d(B_j, B_k) > d_{max} + \varepsilon$, an additional analysis can be performed to determine the cause of overflow and to suggest an appropriate corrective action.

The black-box benchmark difference models can also be organized using various versions of the coefficient of correlation. The *correlation difference model* is the following semimetric:

$$d_{jk} = 1 - r_{jk}^{\alpha} \, , \quad \alpha > 0 \, , \quad (r_{jk} \geq 0) \, .$$

Here r_{jk} denotes the coefficient of correlation between measured performance indicators of benchmarks B_j and B_k, and α is an adjustable parameter whose role is similar to the case of white-box difference. The nominal value is $\alpha = 1$ and we may use $\alpha \neq 1$ only if $r_{jk} \geq 0$. The similarity between benchmarks is defined as $s_{jk} = r_{jk}^{\alpha}$. The fundamental advantage of this approach is that it is invariant with respect to linear transformations. Accordingly, it automatically includes the normalization of data.

In most cases r_{jk} can be interpreted as the coefficient of linear correlation $(r(X, Y) = (\overline{XY} - \overline{X} \cdot \overline{Y})/\sigma_X \sigma_Y)$. In cases where the nonparametric approach is more suitable, r_{jk} can be interpreted as Spearman's rank correlation coefficient:

$$r_s(B_j, B_k) = 1 - \frac{6}{m(m^2 - 1)} \sum_{i=1}^{m} (L_{ji} - L_{ki})^2 \, , \quad -1 \leq r_s(B_j, B_k) \leq +1 \, .$$

where L_{ji} and L_{ki} denote the rank of the ith computer obtained using B_j and B_k respectively. The ranks are positive integers: the rank of the fastest computer is 1, and the rank of the slowest is m.

Both linear correlation and rank correlation reflect the similarity between B_j and B_k: if $0.5 < r(B_j, B_k) \leq 1$ then B_j and B_k are (very) similar, and if $r(B_j, B_k)$ is close to 0 then there is no similarity between B_j and B_k. Let us note that it is possible to have small negative values of $r(B_j, B_k)$ as a consequence of random fluctuations in cases where B_j and B_k are sufficiently different and r is close to 0. However, it is highly unlikely to have large negative values of $r(B_j, B_k)$ because computers (as well as all other engineering products) are designed to have positively correlated performances of individual resources. For example, computer systems with high processor speed typically have large memory and fast disks; it would be counterproductive to combine an increase in processor performance with a decrease in memory capacity and/or file I/O performance. Therefore, it is very unlikely that B_j could produce a ranking that is opposite to the ranking obtained from B_k. In all regular cases the correlation model of difference will generate values in the interval $0 \leq d(B_j, B_k) \leq 1$, yielding the soft limit $d_{max} = 1$. Consequently, the differences in the interval $1 < d(B_j, B_k) \leq 2$ are unlikely but not impossible, yielding for any value of α the hard limit $d_{MAX} = 2$. The soft limit $d_{max} = 1$ will rarely be exceeded, and the hard limit $d_{MAX} = 2$ plays the role of a theoretical upper bound.

From the standpoint of benchmark comparison we can frequently consider B_j and B_k identical if they produce identical rankings of a representative set of m computers. Therefore, the difference metrics can also be based on the difference between rankings. Let R_j and R_k be rankings produced by B_j and B_k. Following is the ranking distance proposed by Kemeny and Snell[15]:

$$
\begin{aligned}
J_{pq} &= 1 & if \quad S_p \succ S_q \quad for \quad B_j \\
&= -1 & if \quad S_p \prec S_q \quad for \quad B_j \\
&= 0 & if \quad S_p \sim S_q \quad for \quad B_j \\
K_{pq} &= 1 & if \quad S_p \succ S_q \quad for \quad B_k \\
&= -1 & if \quad S_p \prec S_q \quad for \quad B_k \\
&= 0 & if \quad S_p \sim S_q \quad for \quad B_k \\
d(R_j, R_k) &= \frac{1}{2} \sum_{p=1}^{m} \sum_{q=1}^{m} |J_{pq} - K_{pq}|
\end{aligned}
$$

The elements of the ordering matrices J and K have the property $J_{pq} = -J_{qp}$ and $K_{pq} = -K_{qp}$. A complete ranking is one containing no ties; e.g. for m=2 such a ranking is $R = (S_1 \succ S_2)$. An opposite ranking is symbolically denoted $-R = (S_1 \prec S_2)$. The "O-ranking" assumes all ties: $O = (S_1 \sim S_2)$. So, for $m = 2$, $d(R, O) = d(-R, O) = 1$ and $d(R, -R) = 2$. Generally, $d(R, O) = d(-R, O) = m(m - 1)/2$ and $d(R, -R) = m(m - 1)$. Using this type of metric, and the concept of limited difference, we can derive the following *ranking difference model*:

$$d(B_j, B_k) = \left(\frac{1}{m(m - 1)} \sum_{p=1}^{m} \sum_{q=1}^{m} |J_{pq} - K_{pq}| \right)^{\alpha}, \quad \alpha > 0.$$

For this model $d(R, O) = d(-R, O) = 1$ and $d(R, -R) = 2^{\alpha}$. So, the soft limit is $d_{max} = 1$ and the hard limit is $d_{MAX} = 2^{\alpha}$.

12.5. CRITERIA FOR BENCHMARK SUITE EVALUATION AND DESIGN

Evaluation and design of benchmark suites is based on a specific set of benchmark suite performance criteria. Two main groups of criteria can be identified: qualitative, and quantitative criteria. We propose the use of ten basic criteria. Qualitative criteria specify global features of benchmark suites and include the following five compliance requirements:

- compliance with the goal of benchmarking
- application area compliance
- workload model compliance
- hardware platform compliance
- software environment compliance

Quantitative criteria are related to the distribution of component benchmark programs in the program space, and include the following five characteristics of benchmark suites:

- size
- completeness
- density
- granularity
- redundancy

Following is a brief description of these ten criteria.

Each benchmark suite supports a specific *goal of benchmarking*. Typical goals are: performance evaluation of a given system, performance comparison of several competitive systems, standardized comparison and ranking of all commercially available systems in a given class, selection of the best system according to specific user requirements, resource consumption measurement and analysis, and performance tuning. A clear and complete specification of the goal of benchmarking is the initial step in all benchmarking efforts. The next step is a convincing justification that a given benchmark suite supports the specified goal. Consequently, the first qualitative criterion benchmark suites must satisfy is the support of a clearly defined goal of benchmarking.

The criterion of *application area compliance* specifies the requirement for a desired application type of benchmark workloads. The benchmark suite members are always assumed to be good representatives of a desired application area. Application areas are sometimes related to the activity of typical users (e.g. scientific, business, educational, and home applications). Another approach is to define application areas according to characteristics of computer workload (e.g. numerical, nonnumerical (combinatorial), seminumerical, graphic, database, systems programming, and networking applications). Each benchmarking effort should be related to a given application area, and so should the members of benchmark suites.

The design of benchmark suites regularly includes the selection of the most appropriate *workload model*. The workload models can be physical, virtual, and functional[12]. The selection of workload model must be justified by specific requirements of a given benchmarking problem. A desired workload model can also be specified during the evaluation of benchmark suites. Once the desired workload model is selected, the benchmark suite must satisfy the workload model compliance criterion.

Functional and virtual workload models are simpler and more frequent than physical models. A recent example of a functional model is a suite of typical business applications for personal computers running under Microsoft Windows, developed by Business Application Performance Corporation[6] (BAPCo). This suite consists of ten popular software products performing typical business functions in the areas of word processing, spreadsheets, database, desktop graphics, desktop presentation, and desktop publishing. As an example of benchmark suites based on a virtual model we can use SPEC benchmarks[1,6]. SPEC92 and SPEC95 are focused on integer and floating point performance as two fundamental components of processor performance. These components can be interpreted as "logical resources" based on performance of several physical resources (processor(s), cache(s), memory, and compilers).

Different functional or virtual workloads can yield very similar usage of hardware and software resources and in such cases their functional/virtual difference becomes insignificant. In such cases we need physical workload models, because they are quantitative and provide measures of redundancy between component programs in a benchmark suite. Generally, there is no justification for using redundant workloads because the cost of benchmarking increases without a corresponding increase in benefits. Therefore, the physical level is fundamental for workload modeling.

The *hardware platform compliance* criterion specifies a target hardware category, and a computer architecture for which a benchmark suite is designed. The most frequent hardware platforms are: personal computers, workstations, mainframes, network servers, local/wide area networks, vector machines, parallel machines, database machines, and communication systems. For each platform it is necessary to identify the main set of resources and to prove that the members of the benchmark suite sufficiently use all identified resources.

The criterion of *software environment compliance* is analogous to the hardware platform compliance criterion. It specifies the desired operating system, windowed environments, programming languages, database systems, communication software, and program development tools that benchmark workloads must use. This criterion identifies all relevant software resources and takes care that they are properly used. In many cases, benchmarking can be limited to the performance analysis of software products (e.g. compilers for a specific language, database systems, operating system overhead, etc.).

Qualitative criteria help to specify a general framework and guidelines for benchmarking efforts. However, all more specific requirements for a set of benchmark workloads must be defined using quantitative criteria. The main advantage of quantitative criteria is the possibility to easily provide a quantitative proof that a benchmark suite satisfies some specific requirements.

The fundamental quantitative criterion is the *size of benchmark suite*, D. It is defined as the diameter of the smallest circumscribed hypersphere containing all benchmark workloads, $B_1, \ldots, B_n$, $n > 1$. The central point of this hypersphere is a hypothetical benchmark B_0 whose coordinates will be denoted $z_{10}, \ldots, z_{m0}$. We can compute $z_{10}, \ldots, z_{m0}$ from the condition that the difference between B_0 and the furthermost of n benchmarks in the suite has the minimum value. Consequently, the size of benchmark suite is

$$D = 2 \min_{z_{10}, \ldots, z_{m0}} \max_{1 \le k \le n} d_{0k} , \qquad d_{0k} = d(B_0, B_k) , \qquad k = 1, \ldots, n .$$

In the case of n benchmark programs, the differences between programs form a symmetric square matrix $[d_{jk}]_{n,n}$, where $d_{jk} = d_{kj}$, and $d_{jj} = 0$. The same properties hold for the similarity matrix $[s_{jk}]_{n,n}$, where $s_{jk} = d_{max} - d_{jk}$. For any group of benchmarks the *central* benchmark, B_c, and the *most peripheral* benchmark, B_p, can be determined from

$$d_{cp} = \min_{1 \le j \le n} \max_{1 \le k \le n} d_{jk} , \qquad s_{cp} = \max_{1 \le j \le n} \min_{1 \le k \le n} s_{jk} .$$

The fundamental property of central benchmark B_c is that its furthermost neighbor, B_p, is located closer than in the case of other benchmarks in the group. So, B_c has the maximum similarity with other benchmarks in the group, and it is usually close to B_0. The central position justifies the use of B_c as the *best representative* of the group. Let us note that the above definition yields B_c and B_p for *any* distribution of workloads, but in some cases the central region of a group of benchmarks can be empty. For example, if we have only two benchmarks, one of them must play the role of B_c, and it is not centrally located. The situation is similar for three benchmarks forming an equilateral triangle. However, real benchmark suites always include centrally located benchmarks which can be used as approximations of B_0. For example, if B_c is centrally located then the difference d_{cp} can be used to compute an approximation of the size of benchmark suite: $D \approx 2d_{cp}$. Of course, this is not a good approximation if B_c is not centrally located.

The maximum value of D corresponds to the simplex distribution of benchmark workloads, where $d(B_j, B_k) = d_{max}$, $j = 1, \ldots, n$, $k = 1, \ldots, n$, $j \ne k$, $n > 1$. Geometrically, such a set of n points in the program space is an equilateral simplex (and no benchmark is centrally located). Since the distance between individual benchmarks has the maximum value d_{max}, the diameter of the circumscribed hypersphere also has the maximum value. The maximum value of D obtained for the simplex distribution depends on n and will be denoted $D_{max}(n)$. This value is the maximum size of a benchmark suite in the case of n benchmarks. Consequently, we can define the following program space *coverage* (or *utilization*) indicator:

$$U(n) = D/D_{max}(n) , \quad 0 < U(n) \leq 1 , \quad n > 1 .$$

The *completeness* of benchmark suites can be evaluated using the coverage indicator $U(n)$. Small values of $U(n)$ indicate incomplete benchmark suites whose members can be excessively redundant, and the values close to 1 indicate benchmark suites that include a wide spectrum of workloads. So, $U \approx 1$ can be considered a necessary condition for a good quality of benchmark suite. Another necessary condition is, of course, a proper distribution of workloads within the program space.

Generally, by increasing the number of benchmark workloads we create an opportunity to cover a spectrum of workload features; this supports increasing the value of n. On the other hand, the cost of benchmarking is proportional to n and this is a reason for reducing the value of n. Hence, the resulting value of n is a compromise between two opposite criteria. To help in selecting the value of n it is useful to define the *density* of a benchmark suite as the number of benchmark workloads per unit of covered program space. Since the coverage is expressed using $U(n)$, the density indicator can be defined as follows:

$$H(n) = n/U(n) \approx (3n - 2)/2D .$$

The concept of granularity is closely related to the concept of density of benchmark suites. We define the benchmark suite *granularity, G*, as the ratio of the number of benchmark programs in the suite, n, and the minimum necessary number of such programs, N, i.e. $G = n/N$. The most reasonable value of N is the number of different computer resources that are used by the benchmark suite. The most frequent resources are: processors, memory, cache memories, disk channels, disks, and software resources. Generally, a computer resource is defined as any major hardware/software component or feature which contributes to the performance of analyzed computer systems.

An obvious reason for designing and using benchmark suites is that a spectrum of performance features of complex hardware/software systems cannot be properly and sufficiently evaluated using a single benchmark workload. If the computer performance is defined as an array of

N performance indicators of individual hardware/software resources, then N benchmark workloads in a benchmark suite play the role of N equations necessary to compute the performance indicators.

In real situations which involve complex and nonlinear phenomena caused by multi-level caching, advanced memory management techniques, parallelism, networking, advanced compiler techniques, etc., it is frequently difficult to clearly identify all relevant hardware and software resources. For example, a control program that is executed by an I/O processor that controls an array of disks can be an important (but not easily visible) resource which in the case of a constrained degree of multiprogramming can limit the global throughput of the file I/O subsystem. Unfortunately, a number of nontrivial experiments is needed to identify the existence and the actual role of such a resource. Consequently, N usually denotes the number of "obvious resources," and the actual number of resources can be greater than N. In such cases the number of benchmarks in the benchmark suite should also be greater than N. Thus, the desirable level of granularity is $G > 1$ This yields a simple guideline for selecting the number of benchmarks in the suite: $n > N$. In the context of the white-box approach N could be interpreted as the dimensionality of the program space. The values of H and G should be as big as possible. However, their maximum values are obviously limited by financial constraints.

Two (or more) benchmark programs are considered *redundant* if their difference is relatively small. Assuming $d_{max} = 100\%$, differences that are less than 15% can usually be considered small differences. In many practical cases it is easy to encounter differences that are less than 5%. In such cases it is difficult to justify why so similar workloads must simultaneously be used because their contribution to the cost of benchmarking is much larger than their contribution to the comparison of competitive systems. The redundancy of benchmark workloads can be visualized using cluster analysis and by presenting benchmark suites in the form of dendrograms[16].

12.6. SIMPLEX BENCHMARK SUITES

A simplex benchmark suite (SBS) is defined as a suite where each pair of benchmarks has the maximum difference d_{max}. This concept is related to the concept of *monoresource benchmarks*[13], where the goal of each component benchmark is to provide maximum utilization of a selected computer resource, and the minimum possible usage of all other resources. Therefore, the simplex suite contains the minimum number of component benchmarks, $n = N$. The main SBS features are:

- a maximum difference between component benchmarks,

- a uniform coverage of the program space,

- the absence of a central benchmark,

- a minimum redundancy between component benchmarks (no benchmark can be removed),

- no benchmark can be added to SBS without introducing redundancy,

- for each number of component benchmarks, SBS has the maximum size and a unit granularity,

- the component benchmarks form an equilateral simplex in the program space.

The goal of SBS is to include component benchmarks which are mutually exclusive and collectively exhaustive. In other words, they should completely cover a spectrum of desired performance characteristics, and should not be redundant. Such programs are *necessary*, and theoretically, *sufficient* for a given performance evaluation task. They are necessary because removing any of programs would result in the total absence of effects caused by some relevant computer resources. They can be sufficient because they contain minimum but complete information about all identified resources. In practice, however, it is not possible to completely isolate the activity of a single resource (e.g. processor activity cannot be eliminated during the measurements of other resources) and it is difficult to prove that all relevant resources are included in SBS.

The concept of SBS is useful both as a theoretical model and as a guideline for practical implementations. The SBS model should be used as a reference point in the design of benchmark suites. Real benchmark suites can be designed as "expanded SBS's" which include SBS component benchmarks plus additional redundant benchmarks which cover the central region of the program space, and provide activity of less important (and initially omitted) computer resources.

Let us now estimate the maximum size of a benchmark suite containing n workloads. The mean absolute difference between a uniformly distributed random variable and the arithmetic mean of n such variables, proved by Milan Merkle[17], is:

$$E|Z_i - (Z_1 + \ldots + Z_n)/n| = \frac{1}{4} - \frac{1}{6n}, \quad Z_i \sim Uniform(0,1), \quad i = 1, \ldots, n.$$

In the case of the uniform random number difference model, uniformly distributed execution times, and approximating B_0 by the arithmetic mean of all benchmarks, we can use the Merkle formula to estimate the maximum size of a benchmark suite and the program space coverage indicator:

$$D_{max}(n) \approx 2 \lim_{m \to +\infty} \frac{3}{m} \sum_{i=1}^{m} |T[i,j] - \frac{1}{n} \sum_{k=1}^{n} T[i,k]| = \frac{3}{2} - \frac{1}{n},$$

$$U(n) \approx \frac{2nD}{3n-2} \approx \frac{4nd_{cp}}{3n-2} = \frac{\min_{1 \le j \le n} \max_{1 \le k \le n} d_{jk}}{0.75 - 1/2n}, \quad 2 \le n \le +\infty.$$

This result can be compared to the case of Euclidean space where for $n = 2$ we start with two points having the distance $D_{max}(2) = d_{max}$. For $n = 3$ the three points form an equilateral triangle inscribed in a circle whose diameter is $D_{max}(3) = 1.15d_{max}$. For $n = 4$ the four points form an equilateral tetrahedron inscribed in a sphere having the diameter $D_{max}(4) = 1.22d_{max}$. Eventually, $D_{max}(+\infty) = \sqrt{2} \cdot d_{max}$. This analysis also shows that in all cases $D_{max}(n)$ is a strictly increasing function of n: $(\forall i > 1) \; D_{max}(i) < D_{max}(i+1)$, but the range is limited: $D_{max}(+\infty) - D_{max}(2) \approx 0.5d_{max}$.

12.7. EVALUATION OF BENCHMARK SUITES

The goal of benchmarking using benchmark suites is to evaluate and compare a given set of hardware/software systems. In the case of standard industrial benchmarking the comparison includes all commercially available computer systems in a given performance range. To evaluate an existing benchmark suite means to analyze the suite using the set of ten basic criteria introduced in the preceding Section.

In order to exemplify the use of basic quantitative criteria for evaluation of benchmark suites let us consider the case of the first SPEC suite

(SPEC89[1]). This example is suitable for analysis because it includes only 10 processor bound benchmarks. These include 4 integer-intensive C programs (*gcc, eqntott, espresso,* and *li*), and 6 floating point intensive FORTRAN programs (*spice2g6, doduc, nasa7, tomcatv, fpppp,* and *matrix300*).

The primary goal of SPEC benchmarks (the first generation, SPEC89, the second generation, SPEC92, and the third generation, SPEC95) is to evaluate integer performance and floating point performance of a processor (or processors) as well as hierarchically organized memory, and to use the resulting compound performance indicators for standard comparison and ranking of commercially available computers with the emphasis on ranking workstations and servers. This type of goal justifies the criterion that two benchmark programs can be considered different if and only if they yield different ranking of evaluated systems. If two benchmark programs produce identical rankings of a set of computers, then, from the standpoint of standardized comparison and ranking, such benchmarks may be considered identical, and their difference (distance) must be zero. The black box difference models are most suitable for the evaluation of benchmark suites of this type.

The goal of the analyzed benchmark suite is restricted to the evaluation of performance of the central processing unit. This goal also restricts the number of performance related computer resources and the corresponding number of necessary component benchmarks. The estimated number of performance related resources is $N = 7$ (the central processor, floating point coprocessor, instruction and data caches, main memory, C compiler and FORTRAN compiler). This yields the acceptable granularity $G = 1.43$, and shows that the numbers of SPEC92 benchmarks (20) and SPEC95 benchmarks (18) can easily be reduced.

In order to generate relative throughput indicators defined in Table 12.1 we used measurements performed by manufacturers of 49 computer systems (the official reports are published in the SPEC Newsletter). Hence, $n = 10$, $m = 49$, and each benchmark is characterized by a 49-component column-vector. These vectors (columns in Table 12.1) are used for computing differences between individual benchmarks. After the normalization of all columns we applied the uniform random number difference model to compute the matrix of differences between benchmarks presented in Table 12.2. The two closest benchmarks are *nasa7* and *tomcatv*: they merely differ by 6.72%. The maximum individual difference is between *matrix300* and *li* (84%). The central

Table 12.2. An analysis of differences between workloads

		1	2	3	4	5	6	7	8	9	10
1	gcc	.00	.18	.25	.24	.55	.21	.12	.68	.48	.54
2	espresso	.18	.00	.32	.35	.66	.13	.15	.78	.64	.64
3	spice2g6	.25	.32	.00	.11	.35	.38	.23	.45	.33	.32
4	doduc	.24	.35	.11	.00	.32	.40	.24	.44	.29	.30
5	nasa7	.55	.66	.35	.32	.00	.72	.55	.15	.10	.07
6	li	.21	.13	.38	.40	.72	.00	.19	.84	.68	.70
7	eqntott	.12	.15	.23	.24	.55	.19	.00	.67	.53	.53
8	matrix300	.68	.78	.45	.44	.15	.84	.67	.00	.20	.14
9	fpppp	.48	.64	.33	.29	.10	.68	.53	.20	.00	.10
10	tomcatv	.54	.64	.32	.30	.07	.70	.53	.14	.10	.00

```
Minimum difference between programs          Dmin =   .07
Average difference between programs          Dave =   .38
Maximum difference between programs          Dmax =   .84
Central program having the min max difference = doduc    (# 4)
Radius of the group (max difference from the CO)   R =   .44
Central object location indicator (<=100%) 100Dmax/2R = 95.00%
```

SPEC89 benchmark is $B_c = doduc$. Its furthermost neighbor $B_p = matrix300$ differs only by 44%, and the size of SPEC89 can be approximated by $2d_{cp} = 0.88$. A more precise result is $D_{SPEC89} = 0.8625$. The maximum size of a 10-benchmark suite is $D_{max}(10) = 3/2 - 1/10 = 1.4$. Thus, the program space coverage (utilization) for SPEC89 is rather modest: $U_{SPEC89} = 0.8625/1.4 = 0.616$ (i.e. less than 62%). The corresponding density is $H_{SPEC89} = 16.2$ workloads per unit of covered program space.

More detailed results of the redundancy and coverage analysis are presented in Fig. 12.1. They include a dendrogram and a covergram generated using a hierarchical clustering technique[16]. The dendrogram is a binary tree whose nodes denote clusters of objects. The nodes are merging points of two clusters (subtrees). The intercluster difference is used to place the node on the workload difference scale (i.e. the distance between the merging point and the left margin equals the difference between the two merging clusters). Each merging point is denoted by the name of the best representative of the cluster.

The most centrally located workload is *doduc* (another centrally located benchmark is *spice2g6*). There are four pairs of closely related benchmarks: {*gcc, eqntott*}, {*espresso, li*}, {*spice2g6, doduc*}, and {*nasa7, tomcatv*}. They differ for less than 13%. This is a high redundancy and indicates both the possibility of reducing the num-

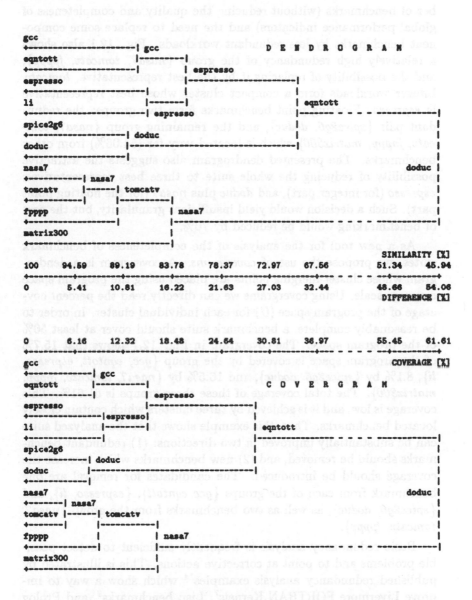

Figure 12.1. SPEC89 redundancy and coverage analysis

ber of benchmarks (without reducing the quality and completeness of global performance indicators) and the need to replace some component benchmarks by less redundant workloads. Fig. 12.1 also shows a relatively high redundancy of the group {nasa7, tomcatv, fpppp}, and the possibility of replacing it by the best representative, tomcatv. Integer workloads form a compact cluster whose best representative is espresso. Floating point benchmarks form two groups: the redundant pair {spice2g6, doduc}, and the remaining group {nasa7, tomcatv, fpppp, matrix300} which is located very far (54.06%) from other benchmarks. The presented dendrogram also suggests the attractive possibility of reducing the whole suite to three best representatives: espresso (for integer part), and doduc plus nasa7 (for the floating point part). Such a decision would yield insufficient granularity, but the cost of benchmarking would be reduced by 70%.

As a new tool for the analysis of the completeness of benchmark suites, we propose the use of covergrams. A covergram is a dendrogram whose cluster merging points are placed using the program space coverage scale. Using covergrams we can directly read the percent coverage of the program space (U) for each individual cluster. In order to be reasonably complete, a benchmark suite should cover at least 50% of the program space. The covergram in Fig. 12.1 shows that 15.7% of the program space is covered by the group {gcc, eqntott, espresso, li}, 8.1% by {spice2g6, doduc}, and 16.5% by {nasa7, tomcatv, fpppp, matrix300}. The total coverage of these three groups is 61.61%. This coverage is low, and it is achieved by three clusters which contain closely located benchmarks. Thus, this example shows that the analyzed suite can be substantially improved in two directions: (1) redundant benchmarks should be removed, and (2) new benchmarks which increase the coverage should be introduced. The candidates for removal are one benchmark from each of the groups {gcc eqntott}, {espresso, li}, and {spice2g6, doduc}, as well as two benchmarks from the group {nasa7, tomcatv, fpppp}.

Basic redundancy analysis is frequently sufficient to detect possible problems and to point at corrective actions. This is illustrated by published redundancy analysis examples[14] which show a way to improve Livermore FORTRAN Kernels[7], Lisp benchmarks[9], and Prolog benchmarks[10]. One conclusion (derived from the analysis of 16 Prolog benchmarks, which all have a maximum intercluster difference of only 9.3%) is that it is useless to apply large benchmark suites in cases where component benchmarks are executed by interpreters. In such cases the only executable in memory is the interpreter itself, and workloads differ only to the extent of variations resulting from executing different parts of the same interpreter. Such variations are typically insignificant (less than 10%), and the whole benchmarking effort can easily be reduced to the use of one to three best representatives of the initial benchmark suite. This result might be a useful hint for designers of Java benchmarks.

The presented evaluation examples illustrate the benefits of the proposed evaluation technique for reducing the redundancy and increasing the coverage of benchmark suites. Consistent use of ten basic evaluation criteria during the design and updates of benchmark suites can improve the majority of their features.

12.8. COVERAGE FUNCTIONS

The most frequent approach to the design of benchmark suites is based on selecting benchmark workloads from a set of existing programs which properly represent the standard activity of an analyzed class of users. A more complex approach is to use (natural or synthetic) workloads with adjustable parameters which affect the utilization of computer resources. In such cases, in addition to selecting a suitable set of workloads, it is also necessary to properly adjust their parameters. This process can always be abstracted as the problem of optimum distribution of points in the program space.

The problem of achieving some desired distribution of workloads in the program space is the fundamental problem of benchmark suite design. In other words, the individual benchmarks are expected to cover the program space in a given way. The coverage of the program space can be analyzed using the concept of *coverage functions*. We assume that each benchmark B_j, $j = 1, \ldots, n$, provides a nonuniform "coverage field" of the program space within a hypersphere Θ_j, with a central point B_j and radius d_{max}. The coverage field denotes the zone of influence created by an individual benchmark. The strongest coverage field is in the central parts of Θ_j, while the coverage field at the circumference and outside of Θ_j must be zero. The total intensity of the coverage field in a given point of the program space is defined as the superposition of intensities of the coverage fields of all individual benchmarks. For simplicity we will assume that the term *coverage* denotes the intensity of the compound coverage field in a given point of the program space.

It may be useful to visualize each benchmark as a source of light that illuminates the interior of the hypersphere Θ_j; of course, central parts are illuminated stronger than the peripheral parts. We can also visualize a set of benchmark workloads as a set of sources of light that are nonuniformly distributed in the program space. The intensity of illumination at each point is obtained by superposition of illumination coming from all sources. The objective of benchmark suite design is to realize a given (uniform or nonuniform) distribution of "illumination" of the program space.

The analogy of the coverage of a program space and the illumination of a physical space is useful to identify the following three ways to achieve the desired distribution:

- by moving (repositioning) workloads in the program space,
- by adding and/or removing selected workloads, and
- by adjusting the "intensity" of each individual workload.

The positioning of individual benchmarks requires adjustable workload parameters, and can be realized only if we use the white-box approach; in such cases we must modify the available benchmark parameters in order to move a program in the direction of the desired destination in the program space. The remaining two techniques can be realized within the black-box approach, and they are discussed in subsequent sections.

Let us now introduce the individual coverage function of the jth benchmark, $\delta_j \mapsto c_j(\delta_j)$, where δ_j denotes the distance from an arbitrary point within Θ_j to the B_j benchmark. We define $c_j(\delta_j)$ as a strictly decreasing function having the following properties:

$$c_j(\delta_j) = \begin{cases} 1, & \delta_j = 0 \\ 0, & \delta_j \geq d_{max} \end{cases},$$

$$\frac{dc_j(\delta_j)}{d\delta_j} < 0, \quad 0 < \delta_j < d_{max}, \quad j = 1, \ldots, n.$$

Let Ω be a given region of the program space. We assume that the global coverage of the program space is obtained by the linear superposition of coverages of all individual benchmarks. Thus, for each point in the region Ω we can define the *compound coverage function*

$$C(\delta_1, \ldots, \delta_n) = \sum_{j=1}^{n} c_j(\delta_j) ,$$

and the *normalized coverage function*

$$C_{norm}(\delta_1, \ldots, \delta_n) = \frac{1}{n} \sum_{j=1}^{n} c_j(\delta_j) .$$

The compound coverage function shows the global intensity of the coverage field resulting from all benchmarks in a benchmark suite. If V denotes the volume of the region Ω then the average compound coverage is

$$C_{ave} = \frac{1}{V} \int_{\Omega} C(\delta_1, \ldots, \delta_n) dV , \quad V = \int_{\Omega} dV .$$

The average compound coverage C_{ave} increases when the density and granularity of a benchmark suite increase, and therefore it can serve as one of the benchmark suite density indicators.

Individual coverage functions can be organized in various ways. First, we define an auxiliary threshold function that has the value 1 inside and the value 0 outside the hypersphere Θ_j:

$$I(\delta_j) = \begin{cases} 1 , & \delta_j < d_{max} \\ 0 , & \delta_j \geq d_{max} . \end{cases}$$

Now we can introduce the following three characteristic forms of the $c_j(\delta_j)$ function:

$$\Lambda(\delta_j) = \left(1 - \frac{\delta_j}{d_{max}}\right) I(\delta_j) = \max\left(0, 1 - \frac{\delta_j}{d_{max}}\right) ,$$

$$\Lambda_1(\delta_j) = \frac{1}{2}\left(1 + \cos\frac{\pi\delta_j}{d_{max}}\right) I(\delta_j) ,$$

$$\Lambda_2(\delta_j) = \frac{1}{2}\left(1 + \cos\frac{\pi(1 - \cos(\pi\delta_j/d_{max}))}{2}\right) I(\delta_j) .$$

These formulas are valid for all values of δ_j, including the rare cases where random fluctuations might cause $\delta_j > d_{max}$. The Λ function yields a linearly decreasing intensity of the coverage field. In the case of Λ_1 and Λ_2 the decrease of the coverage field is nonlinear. Both

Λ_1 and Λ_2 have an intensified coverage in the central part of Θ_j; this effect is particularly strong in the case of Λ_2. Therefore, Λ, Λ_1, and Λ_2 offer three increasing levels of coverage of the central region of Θ_j; their common properties are $c_j(0) = 1$, $c_j(d_{max}/2) = 1/2$, and $c_j(d_{max}) = 0$.

In a general case of n benchmarks and known distances $d_{jk} = d(B_j, B_k)$ we can easily compute the value of the compound coverage function at all points of the program space where benchmarks are located (these points will be called "*benchmark points*"). At the point B_j the distance from B_k is $\delta_k = d_{jk}$ and the compound coverage is

$$C_j = C(d_{j1}, \ldots, d_{jn}) = \sum_{k=1}^{n} c_j(d_{jk}), \quad j = 1, \ldots, n.$$

In the case of the Λ function the compound coverages at benchmark points are

$$C_j = \sum_{k=1}^{n} \max\left(0, 1 - \frac{d_{jk}}{d_{max}}\right) = n - \frac{1}{d_{max}} \sum_{k=1}^{n} d_{jk}, \quad j = 1, \ldots, n.$$

Assuming $d_{max} = 1$ it follows that for all benchmark points, $0 \leq d_{jk} \leq 1$, $1 - d_{jk} = s_{jk}$, and the values of compound coverages reduce to the sum of similarities:

$$C_j = \sum_{k=1}^{n} s_{jk}, \quad j = 1, \ldots, n.$$

The distribution of values C_j, $j = 1, \ldots, n$ is a readily available and easily understandable indicator of the uniformity of coverage of the program space.

12.9. OPTIMUM SUBSETS OF BENCHMARK PROGRAMS

A desired coverage of the program space can frequently be achieved using a suitable subset of the available benchmark programs. The simplest way to create appropriate subsets of benchmark programs is to use the subsets of best representatives created by cluster analysis. The best representatives are the most centrally located benchmarks in selected clusters, and accordingly they cover the program space in an approximately uniform way. For each subset of J benchmarks $(J = n, n - 1, \ldots, 1)$ we can define a binary selector vector $(b_1 \ldots b_n)$, $b_k \in \{0, 1\}$, $k = 1, \ldots, n$ that denotes subsets of benchmark programs: if $b_k = 1$ then B_k is included in a given subset, and if $b_k = 0$ then B_k is excluded. The compound coverage vector at *all* benchmark points, $(C_1 \ldots C_n)$, can now be computed as follows:

$$C_j = \sum_{k=1}^{n} b_k c_j(d_{jk}) , \quad j = 1, \ldots, n .$$

This vector can be compared with the desired distribution of compound coverage at benchmark points. Suppose the desired distribution is uniform. We can determine the average coverage at the benchmark points $\overline{C} = \frac{1}{n} \sum_{j=1}^{n} C_j$, and define the average error of the achieved distribution as the coefficient of variation

$$v_c(C_1, \ldots, C_n) = 100 \left(\frac{n \sum_{j=1}^{n} C_j^2}{\left(\sum_{j=1}^{n} C_j \right)^2} - 1 \right)^{\frac{1}{2}} .$$

The distributions obtained using the subsets of best representatives generated by the cluster analysis can be rather good, but they are *not* the best possible distributions. Indeed, we can find better solutions if we systematically investigate all possible subsets.

Let us again use the binary selector vector $(b_1 \ldots b_n)$ denoting subsets of benchmark programs, and let us introduce the index of the subset, I, the total number of benchmarks in the subset, J, and the uniform distribution error, v_c, as follows:

$$C_j[I; J] = \sum_{k=1}^{n} b_k c_j(d_{jk}) , \quad j = 1, \ldots, n ,$$

$$I = \sum_{k=1}^{n} 2^{k-1} b_k, \ 0 \le I \le I_{max}, \ I_{max} = 2^n - 1; \ J = \sum_{k=1}^{n} b_k, \ 0 \le J \le n,$$

$$v_c[I; J] = v_c(C_1[I; J], C_2[I; J], \ldots, C_n[I; J]) .$$

For each J we can now find the optimum subset of benchmarks:

$$v_c[I^*(J); J] = \min_{1 \le I \le I_{max}} v_c[I; J] , \quad J = 1, \ldots, n .$$

The optimum subsets that contain $1, \ldots, n$ benchmarks and yield the minimum distribution error are denoted by indices $I^*(1), \ldots, I^*(n)$.

In the majority of practical cases the number of benchmarks in a benchmark suite is relatively small (it rarely exceeds 20). Consequently, the minimization of $v_c[I; J]$ can be performed by simply investigating all possible subsets. In the case of larger groups of benchmarks that would not be possible and more sophisticated combinatorial optimization algorithms would be needed.

It is important to emphasize that the quality of distribution is not the only criterion for the design of benchmark suites. Any decrease of the number of individual benchmarks usually decreases both the utilization of program space and the granularity and density of a benchmark suite. Therefore, the selection of optimum subsets of benchmarks can be performed only if we can satisfy the following additional criteria:

- sufficient granularity condition: $G > 1$,

- sufficient density condition: $H > H_{min}$,

- sufficient coverage condition: $U > U_{min}$,

where H_{min} and U_{min} denote the desired threshold values. The design of benchmark suites based on optimum subsets of benchmarks $I^*(1), \ldots, I^*(n)$ consists of selecting a subset that sufficiently satisfies the additional criteria of granularity, density, and coverage.

12.10. OPTIMUM WEIGHTS OF BENCHMARK PROGRAMS

In cases where a benchmark suite already exists it is possible to achieve a desired coverage of the program space by adjusting the individual "intensity" of each benchmark. Let the intensity of the benchmark B_j be expressed using a non-negative real-valued weight W_j. If the Λ coverage model is used, the weighted compound coverage at benchmark points can be defined as follows:

$$C_j = \sum_{k=1}^{n} W_k s_{jk} , \quad W_j \geq 0, \quad j = 1, \ldots, n .$$

Our problem is to determine the weights $W_1, \ldots, W_n$ that will achieve a desired distribution of the compound coverage at points $B_1, \ldots, B_n$.

Let C_j^*, $j = 1, \ldots, n$ be the desired distribution of the compound coverage at the benchmark points. The mean square error function

$$e(W_1, \ldots, W_n) = \sum_{j=1}^{n} \left(C_j^* - \sum_{k=1}^{n} W_k s_{jk} \right)^2$$

in the case of non-singular similarity matrices attains the zero minimum value for weights that satisfy

$$\begin{bmatrix} W_1 \\ W_2 \\ \cdots \\ W_n \end{bmatrix} = \begin{bmatrix} s_{11} & s_{12} & \cdots & s_{1n} \\ s_{21} & s_{22} & \cdots & s_{2n} \\ \cdots & \cdots & \cdots & \cdots \\ s_{n1} & s_{n2} & \cdots & s_{nn} \end{bmatrix}^{-1} \begin{bmatrix} C_1^* \\ C_2^* \\ \cdots \\ C_n^* \end{bmatrix}$$

Unfortunately, the similarity matrix can be singular, or some of the resulting weights can be negative. That is unacceptable and indicates the need to omit the corresponding benchmarks. A more general result can be obtained if we introduce an auxiliary threshold function

$$\mu(W_j) = \begin{cases} 0, & W_j \geq 0 \\ \beta, & W_j < 0, \quad \beta = constant \ (\beta \gg 1) \end{cases}$$

and then define the following compound error function:

$$E(W_1, \ldots, W_n) = \sum_{j=1}^{n} \left(C_j^* - \sum_{k=1}^{n} W_k s_{jk} \right)^2 + \sum_{j=1}^{n} \mu(W_j) .$$

This function can be efficiently minimized using the Nelder-Mead simplex method[17]. The minimization yields non-negative weights that are coordinates of the error function minimum:

$$E(W_1^*, \ldots, W_n^*) = \min E(W_1, \ldots, W_n) , \quad W_j^* \geq 0, \quad j = 1, \ldots, n.$$

The shape of function $E(W_1, \ldots, W_n)$ can be rather irregular and it is important to start minimization from a suitable initial position. The most suitable initial values of weights are $W_j = 1$, $j = 1, \ldots, n$, (i.e. all benchmarks are initially included) and in the important special case of uniform distribution it is suitable to use the constant initial values $C_i^* = \overline{C} = \frac{1}{n} \sum_{j=1}^{n} \sum_{k=1}^{n} s_{jk}$, $i = 1, \ldots, n$.

The resulting weights of component benchmarks $W_1^*, \ldots, W_n^*$ are not normalized (i.e. $W_1^* + \ldots + W_n^* \neq 1$). The normalized (relative) weights are

$$w_j = W_j^* / \sum_{j=1}^{n} W_j^* , \quad j = 1, \ldots, n$$

and can be used to express the relative importance (or "intensity") of each particular benchmark, reflecting the desired distribution of compound coverage at the benchmark points of the program space. For some subset of component benchmarks this method usually yields zero weights ($w_j = 0$, $j \in Z \subset \{1, \ldots, n\}$); such benchmarks are omitted, reducing the cost of benchmarking in all cases which include actual measurements. The results that can be obtained in this way are always better than or equal to the results than can be achieved using the optimum subset method.

The normalized weights can be used for computing the global performance indicators of competitive computer systems. For example, let $t[0, j]/t[i, j]$ denote the relative throughput of the system S_i with respect to the reference system S_0, in the case of the benchmark B_j. Then the average global relative throughput of system S_i can be computed as follows:

$$\overline{R}_i = \prod_{j=1}^{n} \left(\frac{t[0,j]}{t[i,j]} \right)^{w_j} , \quad i = 1, \ldots, m .$$

This performance indicator shows how many times, on the average, system S_i outperforms the reference system S_0 in the case where the comparison is based on the desired distribution of benchmarks in the program space, and the distribution is expressed in terms of the compound coverage at benchmark points.

The described method enables the use of any (sufficiently diversified) set of benchmark programs to achieve an optimized approximation of a desired distribution of the compound coverage. This yields the possibility to design *universal benchmark suites* which uniformly cover the program space, and then to select suitable subsets of these benchmarks in order to express some desired workload properties. Design, maintenance, and updating of such universal suites should be the primary activity of industrial consortia such as SPEC, GPC, and others.

The subsets of benchmarks selected by the optimum weight method must satisfy additional requirements related to granularity, density, and utilization of program space. These requirements can be satisfied either by additional testing of the generated subsets, or by extending the compound error function with additional terms that cause the simplex method to select those optimum weights that simultaneously satisfy all the conditions:

Table 12.3 Compound coverage at benchmark points

COMPOUND COVERAGE (C) AND NORMALIZED COVERAGE (Cn=100*C/N) FOR N = 10 OBJECTS

Object	C	Cn %	0	10	20	30	40	50	60	70	80	90	100%
matrix300	5.67	56.7							*				
li	5.76	57.6							*				
espresso	6.15	61.5							*				
nasa7	6.53	65.3								*			
fpppp	6.65	66.5								*			
tomcatv	6.67	66.7								*			
gcc	6.76	67.6								*			
eqntott	6.80	68.0								*			
spice2g6	7.26	72.6									*		
doduc	7.31	73.1									*		

Average: 6.55   65.5   0   10   20   30   40   50   60   70   80   90   100%
Sigma(compound coverage) = .53   Coefficient of variation = 8.0 %

$$E(W_1,\ldots,W_n) = \sum_{j=1}^{n} \left(C_j^* - \sum_{k=1}^{n} W_k s_{jk} \right)^2 + \sum_{j=1}^{n} \mu(W_j)$$
$$+\mu(G-1) + \mu(H - H_{min}) + \mu(U - U_{min}) \, .$$

12.11. A BENCHMARK SUITE DESIGN EXAMPLE

Let us now design a benchmark suite using the proposed method. Suppose that the suite is primarily intended for the comparison of processing power of Unix workstations. To apply the black-box approach we must initially have a set of candidate workloads representing the activity of a class of users. Suppose that the set of candidate workloads consists of 10 workloads adopted by SPEC in 1989 for their first benchmark suite.

Suppose now that our goal is a uniform coverage of the program space. The average compound coverage at 10 benchmark points using the Λ coverage model is shown in Table 12.3. The distribution is not uniform: the maximum/minimum coverage ratio is 1.29.

The design of a benchmark suite having a uniform distribution of compound coverage at benchmark points yields results shown in Table

12.4. We first present all subsets of best representatives and the corresponding average coverage. The best representatives, however, generate distributions that are not uniform, and the average error with respect to the uniform distribution is frequently greater than 10%.

The distribution of coverage can be improved using the method of optimum subsets. It is interesting to note that a good uniformity of coverage at 10 benchmark points (the error of only 2.3%) can be achieved with only two optimally selected benchmarks (*li* and *matrix300*). When a third benchmark is added the error quickly increases from 2.3% to 7%, and then again decreases when the fourth benchmark is added to the suite. On the other hand, if a small subset of benchmarks can cause a very good coverage at the benchmark points, this might indicate that the benchmark points are located rather closely, and that the redundancy level of benchmarks is too high.

The obtained results can be further improved by using the method of optimum weights. Suppose that we want to achieve the same coverage at all the benchmark points, and that it must be equal to the initial average coverage $\overline{C} = 6.55$. We can achieve this goal with the average error of only 1.563%. This is substantially better than in the case of optimum subsets. The optimum weight method automatically selects the appropriate subset of benchmarks and adjusts their relative importance: the results in Table 12.4 are achieved using only four of the available 10 benchmarks. This reduces the granularity of the suite, but in the case of measuring only the processing power of workstations the granularity for $n = 4$ can still be sufficient. However, the cost of benchmarking is now 2.5 times less than the cost of benchmarking with all candidate workloads.

Therefore, the final result of designing a benchmark suite which uniformly covers the program space and serves for the comparison of processing power of Unix workstations is the suite consisting of four benchmark workloads: *espresso, li, matrix300* and *fpppp*. Their relative weights are respectively 0.126 , 0.38 , 0.452 , and 0.042 . The comparison of m workstations should be performed using the average relative throughputs $\overline{R}_1, \ldots, \overline{R}_m$ computed from the following formula:

$$\overline{R}_i = \left(\frac{t[0,1]}{t[i,1]}\right)^{0.126} \left(\frac{t[0,2]}{t[i,2]}\right)^{0.38} \left(\frac{t[0,3]}{t[i,3]}\right)^{0.452} \left(\frac{t[0,4]}{t[i,4]}\right)^{0.042} .$$

The run times of individual workloads for the reference system are denoted $t[0,1]$, $t[0,2]$, $t[0,3]$, and $t[0,4]$. The reference system can be

Table 12.4. Results of the benchmark suite design

SUBSETS OF BEST REPRESENTATIVES AND THEIR COMPOUND COVERAGE

No.	Average coverage	Average error	Subsets of best representatives
10	6.55	8.0%	1 2 3 4 5 6 7 8 9 10
9	5.89	9.1%	1 2 3 4 5 6 7 8 9
8	5.24	12.0%	1 2 3 4 6 7 8 10
7	4.51	11.8%	1 2 4 6 7 8 10
6	3.83	8.8%	1 2 4 6 8 10
5	3.26	8.0%	1 2 4 8 10
4	2.67	12.5%	1 2 4 5
3	2.00	9.4%	2 4 5
2	1.33	9.2%	5 7
1	.73	17.0%	4

OPTIMUM SUBSETS OF BENCHMARKS AND THEIR COMPOUND COVERAGE

No.	Average coverage	Average error	Optimum subsets of benchmarks
10	6.55	8.0%	1 2 3 4 5 6 7 8 9 10
9	5.82	7.1%	1 2 3 5 6 7 8 9 10
8	5.10	6.2%	1 2 5 6 7 8 9 10
7	4.48	6.6%	1 2 3 5 6 8 10
6	3.76	5.1%	2 5 6 7 8 10
5	3.14	6.0%	2 4 5 6 8
4	2.41	3.2%	2 5 6 8
3	1.87	7.0%	4 6 8
2	1.14	2.3%	6 8
1	.73	17.0%	4

OPTIMUM INTENSITIES OF BENCHMARK PROGRAMS

No.	BENCHMARK	COMPOUND COVERAGE				WEIGHT	
		Initial	Desired	Achieved	Error	Absolute	Relative
1	gcc	6.76	6.55	6.47	-1.3%	.000	.0%
2	espresso	6.15	6.55	6.45	-1.6%	1.423	12.6%
3	spice2g6	7.26	6.55	6.72	2.5%	.000	.0%
4	doduc	7.31	6.55	6.71	2.3%	.000	.0%
5	nasa7	6.53	6.55	6.48	-1.1%	.000	.0%
6	li	5.76	6.55	6.51	-.6%	4.291	38.0%
7	eqntott	6.80	6.55	6.61	.9%	.000	.0%
8	matrix300	5.67	6.55	6.51	-.7%	5.105	45.2%
9	fpppp	6.65	6.55	6.43	-1.9%	.473	4.2%
10	tomcatv	6.67	6.55	6.64	1.3%	.000	.0%

fmin = .105   AVERAGE ERROR = 1.563%

selected as one of the competitive systems. Then $\overline{R}_i$ denotes how many times the ith system is faster than the reference system.

12.12. SUMMARY AND CONCLUSIONS

Any program becomes a benchmark program whenever we focus our interest on its resource consumption instead of its results. However, benchmark workloads and benchmark suites must not be randomly selected. Quantitative methods for evaluation and design of benchmark suites are necessary both for those who create benchmark suites and for those interested in a proper interpretation of performance measurement results. The basic approaches to workload characterization are the white-box and the black-box approach. In cases where it is possible to monitor internal operations and individual resource activities of a computer system, we can use the white-box approach. In cases where the available information is reduced to global performance indicators, such as system response times or throughputs, we use the black-box approach. The theory of program space is applicable in both cases.

The black-box approach can be realized without special measurement tools, and generally it enables easier collection of measurement data than the white-box approach. Therefore, our methodology for evaluation and design of benchmark suites is primarily oriented towards the black-box approach. The proposed evaluation method is based on five qualitative and five quantitative evaluation criteria. The quantitative criteria are used to evaluate the size, redundancy, completeness, granularity, and density of benchmark suites. These criteria are used for both evaluation and design (or upgrading) of benchmark suites. The presented benchmark suite evaluation examples illustrate a general approach which can be used to reduce the redundancy and to improve the completeness of all benchmark suites.

The design of benchmark suites is based on a desired distribution of benchmarks in the program space. In the case of benchmark workloads with adjustable parameters it is possible to move workloads through the program space towards a desired destination. However, this requires the white-box approach, and can differ for various hardware/software architectures. To provide a similar effect within the black-box approach we introduced coverage functions and developed two methods, based on optimum subsets and optimum weights of benchmark programs. The interesting result is that a desired distribution of compound coverage of the program space can be achieved without moving benchmark programs through the program space. If we aggregate the performance measurement results of individual benchmark programs using properly selected weights, then the effects on the aggregate performance indicators are similar to the effects of changing the distribution of benchmarks in the program space.

Thus, it is possible to design universal benchmark suites that uniformly cover the program space. Instead of designing various specific benchmark suites, and performing costly measurements, the desired benchmarking features can be expressed through appropriate weights of individual benchmarks of the universal suite. This technique can reduce the cost of industrial benchmarking and should be widely used for computing aggregate performance indicators, and for comparison of computer systems.

REFERENCES

1. Standard Performance Evaluation Corporation (SPEC), "SPEC Benchmark Suite Release 1.0," SPEC Newsletter 1, 1, pp. 5-9 (Fall 1989).

   This is the first presentation of 10 SPEC89 benchmarks, and the SPECmark performance indicator. This benchmark suite became an industry standard and has been upgraded twice (SPEC92 and SPEC95).

2. W. KOHLER, "How to Improve OLTP Performance and Price/Performance," TPC Quarterly Report, Oct. 15, pp. 1-9 (1992).

3. GPC (Graphics Performance Characterization Committee), "The Effort to Standardize Performance Measurement," GPC Quarterly Report, 2, 4, pp. 10-12 (1992).

4. THE PERFECT CLUB, "The PERFECT Club Benchmarks: Effective Performance Evaluation of Supercomputers," The International J. of Supercomputer Applications, 3, 3, pp. 5-40 (1989).

5. AIM TECHNOLOGY, "The AIM Performance Report: Technical Description," AIM Technology UNIX System Price Performance Guide, pp. 127-129 (Summer 1993).

6. J. GRAY, Ed., The Benchmarking Handbook for Database and Transaction Processing Systems, Morgan Kaufmann, 1991.

This book summarizes all important research results in the area of benchmarking database systems. It is written in a similar style as the first book on benchmarking, N. Benwell (Ed.), Benchmarking - Computer Evaluation and Measurement, J. Wiley, 1975. A "must read" for performance analysts and researchers in this area.

7. F. McMAHON, The Livermore FORTRAN Kernels: a Computer Test of the Numerical Performance Range, Lawrence Livermore National Laboratory, Berkeley, Calif., UCRL-537415, 1986.

8. D.H. BAILEY et al., "The NAS Parallel Benchmarks," The International J. of Supercomputer Applications, 5, 3, pp. 63-73 (1991).

9. R.P. GABRIEL, Performance and Evaluation of Lisp Systems, The MIT Press, Cambridge, Massachusetts, 1986.

A systematic and detailed presentation of a comprehensive suite of Lisp benchmarks. Included are both source programs and the results of performance measurements.

10. L. BURKHOLDER et al., "PROLOG for the People," AI Expert, pp. 63-84 (June 1987).

11. R. GRACE, The Benchmark Book, Prentice-Hall, 1996.

A practitioner's market-oriented survey of currently popular benchmark suites. It includes basic data about SPEC benchmarks, TPC benchmarks, Neal Nelson benchmarks, AIM benchmarks, networking benchmarks, personal computer benchmarks, GPC benchmarks, computer magazine benchmarks, and selected small scientific benchmarks.

12. D. FERRARI, G. SERAZZI, and A. ZEIGNER, Measurement and Tuning of Computer Systems, Prentice-Hall, 1983.

13. J.J. DUJMOVIĆ, "Workload Characterization, Benchmarking, and the Concept of Total Resources Consumption," Proceedings of the Second International Conference on Computer Capacity Management (H.L. Bording and D.J. Schumacher, Ed.), San Francisco, April 8-10, pp. 151-163 (1980).

14. J.J. DUJMOVIĆ, "Clustering, Comparison and Selection of Standard Synthetic Benchmark Programs," Computer Systems Science and

Engineering, **6**, 4, pp. 195-210 (October 1991).

This is the first journal presentation of the black-box approach to workload characterization (the first conference presentation was in June 1988). It includes examples of evaluation of Livermore loops, instruction mixes, Lisp, and Prolog benchmark suites.

15. J.G. KEMENY and J.L. SNELL, Mathematical Models in the Social Sciences, Chapter II (Preference Ranking - An Axiomatic Approach). The MIT Press, 1978.

16. A.K. JAIN and R.C. DUBES, Algorithms for Clustering Data, Prentice Hall, 1988.

17. M. MERKLE, Personal communication, University of Belgrade, 1991.

18. J.A. NELDER and R. MEAD, "A Simplex Method for Function Minimization," The Computer Journal, **7**, pp. 308-313 (1965).

This is a simple, robust, and efficient algorithm for finding a local maximum/minimum of a function of 2 or more variables using only the values of the function (no need for derivatives). The practical value of this method is substantially greater than its popularity. In the Algorithms Hall of Fame the simplex algorithm should be in the first row, together with Quicksort, binary search, and the Runge-Kutta method.

Notes on the Contributors

KALLOL BAGCHI received his MSc in Mathematics from Calcutta University and Ph.D. in Computer Science in 1988 from Jadavpur University, Calcutta, India. He has worked in industry in India and in Finland. He taught a course at the University of Oulu, Finland in 1986. He worked as an **Assistant** and **Associate Professor** in Computer Science and Engineering at Aalborg University, Denmark from 1987-1992. In 1993, he visited Stanford University, CA. He also completed a certificate in Computer Networking from Columbia University, NY. His interests are in performance modeling and simulation, parallel systems. He has authored or co-authored over 40 international papers in these areas. He has been an associate editor or member of the editorial board of the International Journal in Computer Simulation for the last few years and have guest-edited several issues of the journal. He was a member of the board of directors of SCS in 1993. He has been cited in World Who's who and in Who's who in Science and Engineering. He is a member of the ACM and IEEECS. He has been associated with the MASCOTS workshop, since its inception. At present, he is **pursuing a second Ph.D. degree** in business (DIS) at Florida Atlantic University.

RAJIVE BAGRODIA is an **Associate Professor** of Computer Science in the School of Engineering and Applied Science at UCLA. He obtained a Bachelor of Technology in Electrical Engineering from the Indian Institute of Technology, Bombay, in 1981. He obtained his Ph.D. in Computer Science from the University of Texas at Austin in 1987. His research interests include parallel simulation, parallel languages and compilers, and programming methodology. He has published over fifty research papers in refereed journals and international conferences on the preceding topics. He served as the Program Chair and General Chair, respectively, for the 1993 and 1994 Workshop on Parallel and Distributed Simulation, which is jointly sponsored by the IEEE and ACM. He was selected as a 1991 Presidential Young Investigator by NSF. He is also the recipient of the 1992 TRW Outstanding Young Teacher award from the School of Engineering and Applied Science at UCLA and the 1991 Excellence in Teaching award from the Computer Science Department.

AZZEDINE BOUKERCHE is a **Visiting Assistant Professor** in the Department of Computer Sciences at the University of North Texas. He received his M.Sc. and Ph.D., both in Computer Science, from the School of Computer Science at McGill University (Canada). He was employed as a Faculty Lecturer at the School of Computer Science (McGill Univ.) from 1993 to 1995. he also taught at Polytechnic of Montreal during the academic year 1993-1994. He spent the 1991-1992 academic year at the California Institute of Technology where he contributed to a project centered about the specification and verification of the software used to control interplanetary spacecraft operated by JPL laboratory. His current research interests include the design and analysis of methods for parallelizing discrete-event simulations, cellular mobile computing, multiprocessor interconnection networks, and load balancing techniques in distributed systems. His general area of interest is in parallel computing and distributed algorithms in particular. He is the Guest Editor of the Special Issue on Current Advances in Parallel and Distributed Simulation for the International Journal of Computer Simulation. Dr. Boukerche is a member of the IEEE Computer Society and the ACM.

GIANFRANCO CIARDO is an **Assistant Professor** in the Department of Computer Science at the College of William and Mary, Williamsburg, Virginia. He is interested in theory and tools for the behavioral, performance, and reliability modeling of complex hardware/software systems, with an emphasis on stochastic Petri nets. In this area, he has published over 30 refereed journal and conference papers, including several at the Petri Nets and Performance Models (PNPM) cycle of workshops, for which he served as co-chair in 1995. He is a designer and developer of SPNP, the Stochastic Petri Net Package, in use at some 50 sites worldwide, and has consulted for both industry and government.

SAJAL K. DAS is an **Associate Professor** of Computer Sciences and also the Director of the Center for Research in Parallel and Distributed Computing at the University of North Texas, Denton. He is a recipient of Cambridge Nehru Fellowship and an Honor Professor award from UNT. His current research interests include the design and analysis of parallel algorithms and data structures, parallel discrete-event simulation, cellular mobile computing, multiprocessor interconnection networks, and load balancing techniques in distributed systems. Das has published over 75 technical papers and presented his research in numerous conferences.He is an Editor of Parallel Processing Letters, Journal of Parallel Algorithms and Applications, and the CD-ROM Journal of Computing and Information. He has served on the Program Committees of several conferences. Das is a member of the IEEE Computer Society and the ACM.

JOZO J. DUJMOVIĆ is Professor of Computer Science at San Francisco State University. His previous faculty positions were at Worcester Polytechnic Institute, the University of Texas at Dallas, the University of Florida,Gainesville, and the University of Belgrade, where he was Chairman of the Computer Science Department in the School of Electrical Engineering. He received the Dipl. Ing. degree in Electrical Engineering, and the Master of Science and the Doctor of Science degrees in Computer Science, all from the University of Belgrade, Yugoslavia. His research interests are in computer performance evaluation, software engineering, and preferential neural networks. He published more than 70 papers and 8 books, and worked for industry as a computer hardware designer, system software designer, and evaluator of complex systems.

ZULAH K. F. ECKERT earned her Ph.D. from the Department of Computer Science in 1995. She is interested in programming languages, including parallel and distributed languages, program nondeterminism, and parallel execution constructs. Her interests also include performance analysis tools, computer simulation, distributed shared memory systems, computer networks, visualization, and visual education tools. She enjoys teaching and has constructed visualization tools to aid in teaching computer science concepts including formal languages, program verification, and data structures. She is a **Consultant** designing and teaching industrial education courses, currently in the area of computer networks.

A. J. (TONY) FIELD graduated from the University of Reading in 1981, winning one of the University Prizes for his BSc. He completed his PhD at Imperial College in the design and performance analysis of switching networks in 1985 and became a **Lecturer** in Computing in 1986. Since then he has worked predominantly in parallel systems, focussing on performance modelling of interconnection networks and large-scale distributed cache memory systems. He has published one book and over twenty research papers in the areas of functional programming, computer architecture and performance modelling. He has also held a number of research grants and was made a Teaching Fellow of Imperial College in 1995.

PETER HARRISON is currently a **Reader** in Computing Science at Imperial College where he became a lecturer in 1983. He graduated at

Christ's College Cambridge as a Wrangler in Mathematics in 1972 and went on to gain Distinction in Part III of the Mathematical Tripos in 1973, winning the Mayhew prize. He obtained his PhD in Computing Science at Imperial College in 1979. He has researched into analytical performance modelling techniques and algebraic program transformation for some fifteen years, visiting IBM Research Centers for two summers in the last decade. He has written two books, had over 70 research papers published in the areas of performance modelling and functional programming and held a series of research grants.

JANE HILLSTON received a B.A. degree in Mathematics from the University of York, England in 1985, and an M.Sc. degree in Mathematics from Lehigh University, USA in 1986. She completed her Ph.D. degree in Computer Science at the University of Edinburgh in 1994. After a brief spell working in industry, she moved to the Department of Computer Science at the University of Edinburgh in 1989 as a research associate. In 1995 she was appointed a **Lecturer** in the same department. Dr Hillston's current research interests include process algebras, Markov processes, stochastic Petri nets, and aggregation and decomposition techniques.

FRED HOWELL completed his BSc MEng in Microelectronic Systems Engineering at U.M.I.S.T. in 1992. He spent the summers working for ICL in Manchester on hardware and software design for the Flagship and EDS parallel machines. He has since spent brief spells working for PC Magazine and Digital (Scotland) Ltd. He is currently in the Department of Computer Science at the University of Edinburgh **completing a PhD** in parallel program design techniques and working on an EPSRC project using simulation to evaluate multiprocessor interconnection networks.

ROLAND N. IBBETT (PhD, FRSE, C.Eng, FBCS, FRSA) was appointed as Lecturer in Computer Science at theUniversity of Manchester in 1967 and was subsequently promoted to Senior Lecturer and then Reader. In 1985 he was appointed to a **Chair in Computer Science** at the University of Edinburgh. His research interests in computer architecture have led to the publication of over 40 papers and two books. While at Manchester he was a major contributor to the design of the MU5 computer. His use of the MU5 logic simulator, and involvement in the use of ISPS during a semester at Carnegie-Mellon University, identified a need for a higher level architecture simulator and thus led to the design of HASE.

VIJAY K. MADISETTI obtained his Ph.D in Electrical Engineering and Computer Sciences at the University of California at Berkeley in 1989. He is an **Associate Professor** of Electrical and Computer Engineering at Georgia Tech. He has authored an IEEE Press textbook *VLSI Digital Signal Processors*, 1995. He is also the president of VP Technologies, Inc, specializing in virtual prototyping and electronics system design. Madisetti may be contacted at *vkm@ee.gatech.edu* on the Internet.

BRIAN A. MALLOY is an **Assistant Professor** in the department of Computer Science at Clemson University. Prior to Clemson, he was a lecturer at the University of Pittsburgh and an assistant professor at Duquesne University. Dr. Malloy's research extends to several areas of computer science. He has developed techniques for compiler optimization including instruction scheduling, parallelization and cache optimization. He has developed program representations to facilitate analysis and optimization; these representations include program dependence graphs for imperative and object-oriented software. He has developed several languages for process-oriented simulation including SimCal, for Pascal, and SimPol, for C++. In addition, he has have developed techniques for parallelizing discrete event simulation.

GARY J. NUTT is the **Chair and a Professor** in the Department of Computer Science at the University of Colorado at Boulder. He is interested in many aspects of distributed computing, including operating systems, collaboration technology, performance evaluation, modeling, visualization, networks, and the effective transfer of these technologies into environments where they can be used to solve people's problems. He has worked at research labs and on real products in industry as well as having been at the university for almost twenty years. He earned his Ph.D. from the University of Washington in 1972.

MARINA RIBAUDO graduated in Computer Science at the University of Torino (Italy) in 1990 and she was awarded a Ph.D. in Computer Science from the same University in 1995. Since July 1995 she has been a **Researcher** at the Computer Science Department of the University of Torino. Dr Ribaudo's current research interests are in the area of performance evaluation of computer systems, mainly in the fields of stochastic Petri nets and stochastic process algebras. Her Ph.D thesis examines the relationships existing between these two formalisms.

MOHAMED S. BEN ROMDHANE obtained his Ph.D. in Electrical Engineering at Georgia Institute of Technology in 1995. He is currently a **member of the technical staff** of Rockwell International Corporation. He has authored a Kluwer Academic P. book *Quick-Turnaround ASIC Design in VHDL --Macrocell Based Synthesis*, 1996. His current research interests are in hardware/software co-design for digital signal processing and communications systems. Ben Romdhane can be contacted at *romdhane@risc.rockwell.com* on the Internet.

DIANE T. ROVER is currently an **Assistant Professor** in the Department of Electrical Engineering at Michigan State University. Under a DOE Postdoctoral Fellowship from 1989 to 1991, she was a research staff member in the Scalable Computing Laboratory at the Ames Laboratory. She has held summer positions with McDonnell Douglas Corp. and the IBM Thomas J. Watson Research Center. Her research interests include integrated program development and performance environments for parallel and distributed systems, instrumentation systems, performance visualization, embedded real-time system analysis, and reconfigurable hardware. She received the B.S. degree in Computer Science in 1984, the M.S. degree in Computer Engineering in 1986, and the Ph.D. degree in Computer Engineering in 1989, all from Iowa State University. From 1985 to 1988, she was awarded an IBM Graduate Fellowship. She has received an R&D 100 Award in 1991 for the development of the Slalom benchmark, a MasPar Challenge Award in 1994, an MSU College of Engineering Withrow Teaching Excellence Award in 1994, and an NSF CAREER Award in 1996. Dr. Rover is a member of the IEEE Computer Society, ACM, and ASEE, and is currently the Director of Technical Activities for the IEEE Southeastern Michigan Section.

RAFAEL H. SAAVEDRA is an **Assistant Professor** in the Computer Science Dept. at the University of Southern California. He received his Ph.D. from U.C. Berkeley in 1992.

ALAN JAY SMITH was raised in New Rochelle, New York, USA He received the B.S. degree in electrical engineering from the Massachusetts Institute of Technology, Cambridge, Massachusetts, and the M.S. and Ph.D. degrees in computer science from Stanford University, Stanford, California. He is currently a **Professor** in the Computer Science Division of the Department of Electrical Engineering and Computer Sciences, University of California, Berkeley, California, USA His research interests include the analysis and modeling of computer systems and devices, computer architecture, and operating systems.

Dr. Smith is a Fellow of the Institute of Electrical and Electronic Engineers, and is a member of the Association for Computing Machinery, IFIP Working Group 7.3, the Computer Measurement Group, Eta Kappa Nu, Tau Beta Pi and Sigma Xi. He is on the Board of Directors (1993-95), and was Chairman (1991-93) of the ACM Special Interest Group on Computer Architecture (SIGARCH), was Chairman (1983-87) of the ACM Special Interest Group on Operating Systems (SIGOPS), was on the Board of Directors (1985-89) of the ACM Special Interest Group on Measurement and Evaluation (SIGMETRICS), was an ACM National Lecturer (1985-6) and an IEEE Distinguished Visitor (1986-7), was an Associate Editor of the ACM Transactions on Computer Systems (TOCS) (1982-93), is a subject area editor of the Journal of Parallel and Distributed Computing and is on the editorial board of the Journal of Microprocessors and Microsystems. He was program chairman for the Sigmetrics '89 / Performance '89 Conference, program co-chair for the Second (1990) and Sixth (1994) Hot Chips Conferences, and has served on numerous program committees.

ABDUL WAHEED is currently a **Ph.D. candidate** in the Department of Electrical Engineerig at Michigan State University. Currently he works as a research assistant for an ARPA-funded project. He has taught several undergraduate courses in the Department of Electrical Engineering as a teaching assistant. He worked with Siemens Medical Engineering Divison in 1991 as a field service engineer. He held a summer position at Hewlett-Packard Labs in 1994 as a software engineer. His research interests include parallel and distributed systems, design, evaluation, and management of instrumentation systems, modeling and simulation of computer systems, and real-time systems. He received the B.Sc. degree in Electrical Engineering from the University of Engineering and Technology, Lahore, Pakistan in 1991 and the M.S. degree in Electrical Engineering from Michigan State University, East Lansing in 1993.

JERRY WALDORF obtained a MS in Computer Science and BS in Computer Science and Engineering from the University of California, Los Angeles (UCLA) in 1993 and 1989, respectively. He is currently working as a **Software Engineer** in language design at the Software Technologies Corporation. He is a member of the ACM.

GEORGE W. ZOBRIST received his BS and PhD in Electrical
Engineering from the University of Missouri-Columbia in 1958 and 1965,
respectively and his MSEE from the University of Wichita in 1961.

He has been employed by industry, government laboratories and various
Universities during his career. He is presently **Chairman/Professor** of
Computer Science at the University of Missouri-Rolla.

His current research interests include: Simulation, Computer Aided
Analysis and Design, Software Engineering and Local Area Network
Design. He is presently Editor of IEEE Potentials Magazine, VLSI Design
and International Journal in Computer Simulation.

Author Index

Subject Index

Abstract performance machine
  model, 258
Animation, 11
Artificial rollback, 117

Benchmark suite design criteria,
  297
Benchmark suite:
  best representative, 300
  completeness, 301
  covergram, 307
  definition, 288
  dendrogram, 307
  density, 301
  evaluation criteria, 297
  granularity, 301
  optimum subsets of
    programs, 313
  optimum weights of
    programs, 315
  redundancy, 302
  simplex suite, 303
  size, 300
Black-box workload difference
  models, 292
BoNeS, 11
Boundary virtual time, 150

C++, 2
Cache coherency, 57
Cache line states, 64, 75
Cancelation strategies, 119
Cancelback, 116
Causality constraint, 108
Chernoff Faces, 273
Coherency operations, 65
Computational model, 20, 21, 25,

31, 32
Conservative parallel simulation,
  135, 139
Conservative mechanisms, 109
Continuous-time Markov chain,
  211
Crossbar network, 14

Deadlock, 109
DEMOS, 10
Discrete-time Markov chain, 216,
  217
Distributed memory, 118
Distributed simulation *see* parallel
  simulation
DORMS, xx

Equivalence relations, 246-250
Error propagation, 120
Event trace, 93
Execution time prediction, 276
Execution-driven simulation, 20-
  23, 28, 30-32
Fossil collection (FC), 115

Global virtual time (GVT), 111,
  136, 149

Instrumentation system, 35, 37
Integrated environment, 40
IS characterization, 49
IS evaluation, 46
IS management, 43

Local virtual time, 149

Machine characterizer, 265

T - #0057 - 101024 - C0 - 229/152/20 [22] - CB - 9789056995690 - Gloss Lamination